Mechanical Alloying: Processing and Materials

Mechanical Alloying: Processing and Materials

Editor

Joan-Josep Suñol

MDPI • Basel • Beijing • Wuhan • Barcelona • Belgrade • Manchester • Tokyo • Cluj • Tianjin

Editor
Joan-Josep Suñol
Universitat de Girona
Spain

Editorial Office
MDPI
St. Alban-Anlage 66
4052 Basel, Switzerland

This is a reprint of articles from the Special Issue published online in the open access journal *Metals* (ISSN 2075-4701) (available at: https://www.mdpi.com/journal/sustainability/special_issues/Bioenergy_Biofuels).

For citation purposes, cite each article independently as indicated on the article page online and as indicated below:

LastName, A.A.; LastName, B.B.; LastName, C.C. Article Title. *Journal Name* **Year**, *Volume Number*, Page Range.

ISBN 978-3-0365-2117-6 (Hbk)
ISBN 978-3-0365-2118-3 (PDF)

Contents

About the Editor

Joan-Josep Suñol is full professor in Applied Physics at the University of Girona, Spain (since 2020). He works in the Department of Physics of the Polytechnic Engineering School. He obtained the Ph.D. degree at the Autonomous University of Barcelona in 1996. Regarding publications and congresses, more than 200 articles are indexed (SCI) journals and more than 300 communications has been published in scientific congress. Dr. Suñol has been the president of the Spanish GECAT (Thermal Analysis and Calorimetry Group) since 2015 and the coordinator of the Materials and Thermodynamics research group of the University of Girona. He was head of the Physics Department of the University of Girona from 2007 to 2010.

Preface to "Mechanical Alloying: Processing and Materials"

This book is a compilation of recent articles linked to the production and the structural and functional characterization of alloys and compounds produced by mechanical alloying. In one of the works in this Special Issue, a non-complex model of mechanical alloying was applied to compare the final microstructure of two nanocrystalline alloys as a function of the energy transfer in two milling devices: planetary and shaker. Regarding the production of materials, some examples are: (a) milled Mn–Al-based alloys were introduced in dissolutions with azoic dye (wastewater treatment redox processes); (b) controlled annealing provoked the relaxation of mechanically induced strain or the recrystallization from an amorphous phase.

Mechanical alloying can be considered as a step-in powder metallurgy process (high-temperature press, spark plasma, or microwave sintering). Usually, the main objective is to obtain the desired microstructure to optimize the mechanical and functional properties of the material.

As Guest Editor of this Special Issue, I am very happy with the final result, and hope that the selected papers will be useful to researchers working on mechanical alloying as a processing technique of materials with improved functional properties. I would like to warmly thank the authors of the eight articles in this Special Issue for their contributions, and all of the reviewers for their efforts in ensuring high-quality publications. Finally, thanks to the editors of Metals for their continuous help, and to the Metals editorial assistants for their valuable and inexhaustible engagement and support during the preparation of this volume.

Joan-Josep Suñol
Editor

metals MDPI

Editorial

Mechanical Alloying: Processing and Materials

Joan-Josep Suñol

Department of Physics, C/Universitat de Girona 3, Universitat de Girona, 17003 Girona, Spain; joanjosep.sunyol@udg.edu; Tel.: +34-972-419-757

1. Introduction and Scope

Mechanical alloying is a technique involving the production of alloys and compounds, which permits the development of metastable materials (with amorphous or nanocrystalline microstructure) or the obtention of solid solutions with extended solubility. The elements or compounds to be mix (usually as powders) were introduced in jars, together with a few numbers of balls.

Regarding the scope of this Special Issue, so many options were given to the potential authors:

1. Synthesis and processing in solid-state science and technology: high-energy milling, severe plastic deformation of materials (SPD), reaction milling.
2. New materials/processes: oxide dispersion strengthened (ODS) alloys, nanomaterial, nano-composites, and quasi-crystalline phases/materials.
3. Structural characterization: mechanically induced structural changes in materials (point defects, dislocations, clusters, precipitates, grain boundaries), surfaces and interfaces in activated solids.
4. New equipment and procedures: milling equipment based on improved milling dynamics, processing optimization and milling contamination.

Finally, only height articles have been published. Nevertheless, the set of materials, characterization and applications described in the manuscripts provides a wide spectrum of the potential of this processing technique.

check for updates

Citation: Suñol, J.-J. Mechanical Alloying: Processing and Materials. *Metals* **2021**, *11*, 798. https://doi.org/10.3390/met11050798

Received: 6 May 2021
Accepted: 13 May 2021
Published: 14 May 2021

2. Contributions

Regarding the modelling of the milling process, the main problem is due to the high quantity of processing parameters to be controlled, which include the filling factor of the jars, the material of the jars and balls, the milling atmosphere, the milling time, the milling intensity, the ball to powder weight ratio (BPR), the number and diameter of balls, the temperature inside the jars, the local temperature on interactions between powder and balls, the optional change in the sense of the rotation of the jars, the on–off switch periods, the controlled addition of a process control agent (PCA) that can help in grain refinement and act as a surfactant, the frequency of collisions between balls, in which powdered particles are involved, and so on. Thus, it is quite difficult to model the energy or powder transfer during the milling process. Furthermore, there are ball milling devices with different geometries: shaker mills, planetary mills. Likewise, the interaction between of the powders with balls (and/or jar internal wall) can be facilitated by abrasion or percussion. For kinetic energy, the velocity of the balls has a broad distribution. For this, all models are usually based upon estimation. One of the works in this Special Issue applies a non-complex model to compare the final microstructure of two Fe-X-Nb-Cu (X = Nb, Ni-Zr) alloys as a function of the energy transfer in two milling devices: planetary and shaker. In this work, the shaker mill is more energetic [1].

Regarding the production of materials, the alloys and compounds that are produced are obtained in a powder shape. Milling usually favors a reduction in the grain size (except for very ductile materials) and the formation of smooth surfaces with high specific surface/volume ratio. The size distribution of the powders can be checked by scanning electron microscopy. One of the problems associated with the milling process is contamination from the milling tools and atmosphere. Additional oxygen contamination can be induced after the extraction of the powders from the jars. Thus, a shift in the composition can be produced. This effect is checked with microanalysis techniques.

Sometimes the powdered compounds can be directly used in specific applications without additional treatments. As an example, Mn–Al-based alloys were introduced in dissolutions with azoic dye. The interaction with the metallic particles favors the decolorization process of the dyes by breaking the azo bond of the macromolecule [2].

In order to obtain an improvement in the functional properties of the alloys and compounds, sometimes controlled annealing is needed. Furthermore, annealing provokes the relaxation of the mechanical induced strain. In one of the articles in this Special Issue, annealing was performed at 700–800 °C in high-nitrogen chromium-manganese steels [3]. The austenite phase of the steel was stabilized. Likewise, the annealing objective is the re-crystallization of an amorphous phase [4]. The development of the desired crystallographic phase is associated with the influence of the microstructure in the functional response of the alloy. Some Mn-Co(Fe)-Ge(Si) alloys have a martensitic transformation coupled with a magnetic transition favoring an improved magnetocaloric effect.

Mechanical alloying can be a step-in powder metallurgy process. The powders (as obtained after milling) can be compacted at high pressure. An alternative is the spark plasma sintering process (SPS). Refractory high-entropy alloys are produced to maximize the strength, yield strength and fracture strain [5]. An innovative technique is the microwave sintering of previously compacted powders [6]. Al-Y_2O_3 nanocomposites produced by mechanical alloying and pressing were sintered in a microwave sintering oven. The processing conditions were heating rate of 10 K/min until 550 °C and a dwell time of 30 min. The main objective is to optimize the mechanical properties: hardness, yield strength, ultimate compression strength and compressive strain.

Two of the selected articles are reviews. One is devoted to Fe-Cr based alloys and their consolidation at high temperature [7]. In these materials, the technological objective is to improve the resistance to corrosion. Nanocrystalline alloys have higher resistance than microcrystalline alloys. A system with improved resistance is Fe-Cr-Ni-Zr.

A second article revises the hydrogen absorption behavior and the absorption/desorption kinetics of metal hydrides produced by mechanical alloying [8]. It is a critical overview on the effect of mechanical alloying in binary (CaH_2, MgH_2, etc.) and ternary (Ti-Mn-N and Ca-La-Mg-based systems) hydrides. Sometimes the technological process has multiple steps, involving: mechanical alloying, heat treatment, a second mechanical alloying process, degassing and, finally, extrusion.

3. Conclusions and Outlook

As a main conclusion, it is necessary to acknowledge the variety of alloys and compounds produced by mechanical alloying: Fe-X-B-Cu (X = Nb, NiZr) nanocrystalline alloys, mixtures of the binary Fe-Mn and Fe-Cr alloys with the nitrides CrN (Cr_2N) and Mn_2N, Mn-Al-Co and Mn-Al-Fe alloys, non-equiatomic refractory high entropy alloy $(W_{35}Ta_{35}Mo_{15}Nb_{15})_{95}Ni_5$, nanocrystalline $MnCo_{0.8}Fe_{0.2}Ge_{1-x}Si_x$, nanocrystalline Fe-Cr alloys, Al–Y_2O_3 nanocomposites and hydride-forming alloys. Regarding the study of their properties, it is important to improve mechanical properties, hydrogen absorption, magnetocaloric effect and resistance to corrosion. The processing parameters affect the final microstructure of the material, and the microstructure affects the functional response. Likewise, the powders can be consolidated (press, spark plasma sintering, microwave sintering) to obtain bulk materials. Further investigations should be performed to gain a deeper knowledge of the influence of the milling parameters and to analyze the option to develop new advanced materials for specific applications.

As Guest Editor of this Special Issue, I am very happy with the final result, and hope that the present selected papers will be useful to researchers working on mechanical alloying as processing technique of materials with improved functional properties. I would like to warmly thank the authors of the eight articles in this Special Issue for their contributions, and all of the reviewers for their efforts in ensuring high-quality publications. Finally, thanks also to the editors of *Metals* for their continuous help, and to the *Metals* editorial assistants for the valuable and inexhaustible engagement and support during the preparation of this volume. In particular, my sincere thanks go to Toliver Guo for his help and support.

Conflicts of Interest: The author declares no conflict of interest.

References

1. Carrillo, A.; Saurina, J.; Escoda, L.; Suñol, J.J. Fe-X-B-Cu (X = Nb, NiZr) alloys produced by mechanical alloying: Influence of milling device. *Metals* **2021**, *11*, 379. [CrossRef]
2. Ben Mbarek, W.; Saurina, J.; Escoda, L.; Pineda, E.; Khitouni, M.; Suñol, J.J. Effects of the addition of Fe, Co on the azo dye degradation ability of Mn-Al mechanically alloyed powders. *Metals* **2021**, *10*, 1578. [CrossRef]
3. Lyashkov, K.; Shabashovm, V.; Zamatovskii, A.; Koslov, K.; Kataeva, N.; Novikov, E.; Ustyugov, Y. Structure-phase transformation in the course of solid-state mechanical alloying of high-nitrogen chromium-manganese steels. *Metals* **2021**, *11*, 301. [CrossRef]
4. Vidal-Crespo, A.; Ipus, J.J.; Blázquez, J.S.; Conde, A. Mechanical amorphization and recrystallization of Mn-Co(Fe)-Ge(Si) compositions. *Metals* **2019**, *9*, 534. [CrossRef]
5. Duan, B.; Yu, Y.; Liu, X.; Wang, D.; Wu, Z. A Novel Non-Equiatomic $(W_{35}Ta_{35}Mo_{15}Nb_{15})_{95}Ni_5$ Refractory High Entropy Alloy with High Density Fabricated by Powder Metallurgical Process. *Metals* **2020**, *10*, 1436. [CrossRef]
6. Matli, M.R.; Shakoor, R.A.; Matli, P.R.; Mohamed, A.M.A. Microstructure and Compressive Behavior of Al–Y_2O_3 Nanocomposites Prepared by Microwave-Assisted Mechanical Alloying. *Metals* **2019**, *9*, 414. [CrossRef]
7. Singh Raman, R.K. Mechanical alloying of elemental powders into nanocrystalline (NC) Fe-Cr alloys: Remarkable oxidation resistance of NC alloys. *Metals* **2021**, *11*, 695. [CrossRef]
8. Somo, T.R.; Maponya, T.C.; Davids, M.W.; Hato, M.J.; Lototskvy, M.V.; Mobidane, K.D. A Comprehensive Review on Hydrogen Absorption Behaviour of Metal Alloys Prepared through Mechanical Alloying. *Metals* **2020**, *10*, 562. [CrossRef]

metals

Article

Microstructure and Compressive Behavior of Al–Y_2O_3 Nanocomposites Prepared by Microwave-Assisted Mechanical Alloying

Manohar Reddy Mattli [1], R. A. Shakoor [1,*], Penchal Reddy Matli [2] and Adel Mohamed Amer Mohamed [3]

1 Center for Advanced Materials, Qatar University, Doha 2713, Qatar; manoharreddy892@gmail.com
2 Department of Mechanical Engineering, National University of Singapore, Singapore 117576, Singapore; drlpenchal@nus.edu.sg
3 Department of Metallurgical and Materials Engineering, Faculty of Petroleum and Mining Engineering, Suez University, Suez 43721, Egypt; adel.mohamed25@yahoo.com
* Correspondence: shakoor@qu.edu.qa; Tel.: +974-44036867

Received: 4 March 2019; Accepted: 3 April 2019; Published: 5 April 2019

Abstract: In this study, Al–Y_2O_3 nanocomposites were synthesized via mechanical alloying and microwave-assisted sintering. The effect of different levels of yttrium oxide on the microstructural and mechanical properties of the Al–Y_2O_3 nanocomposites were investigated. The density of the Al–Y_2O_3 nanocomposites increased with increasing Y_2O_3 volume fraction in the aluminum matrix, while the porosity decreased. Scanning electron microscopy analysis of the nanocomposites showed the homogeneous distribution of the Y_2O_3 nanoparticles in the aluminum matrix. X-ray diffraction analysis revealed the presence of yttria particles in the Al matrix. The mechanical properties of the Al–Y_2O_3 nanocomposites increased as the addition of yttria reached to 1.5 vol. % and thereafter decreased. The microhardness first increased from 38 Hv to 81 Hv, and then decreased to 74 ± 4 Hv for 1.5 vol. % yttria. The Al–1.5 vol. % Y_2O_3 nanocomposite exhibited the best ultimate compressive strength and yielded a strength of 359 ± 7 and 111 ± 5 MPa, respectively. The Al–Y_2O_3 nanocomposites showed higher hardness, yield strength, and compressive strength than the microwave-assisted mechanically alloyed pure Al.

Keywords: aluminum; yttrium oxide (yttria); mechanical alloying; microwave sintering; microstructure and mechanical properties

1. Introduction

Metal matrix composites (MMCs) find noteworthy applications in many engineering sectors due to their superior properties such as high strength, high-temperature capability, specific modulus, and good wear resistance compared to monolithic base materials. The mechanical performances of MMCs often show greater improvement than can be achieved by conventional strengthening methods in monolithic alloys [1–4].

Aluminum (Al)-based metal matrix composites (AMMCs) are an excellent choice for automotive, aerospace, defense, and nuclear power sectors because of their lightweight and favorable mechanical, thermal, and physical properties. Aluminum (Al)-based metal matrix composites are capable of achieving high strength, high-fatigue resistance, high-wear and corrosion resistance, and good compatibility with various manufacturing processes [5–8].

At present, ceramic particle-reinforced Al-matrix nanocomposites have been prepared primarily by mechanical alloying, forging, and casting routes [9–11]. Among these methods, mechanical alloying (MA) has been widely used to fabricate Al-matrix nanocomposites due its cost-effectiveness, simplicity, and its ability to improve the properties vis-a-vis those of the unreinforced matrix [12,13]. There are many sintering techniques such as conventional, spark plasma, vacuum, and microwave sintering processes [14–17]. Among these techniques, the microwave sintering process is a heating method that offers the ability to balance the radiant and microwave heating effects. In this process, heat is generated within the sample by rapid oscillation of dipoles at microwave frequencies. Microwave sintering provides efficient internal heating, and energy is supplied directly to the material. Therefore, this process avoids the significant temperature gradient between the surface and interior. Microwave sintering is a high-technology heating process that can save both energy and time [18].

In AMMCs, the most common types of reinforcement that can be used are SiC, Si_3N_4, Y_2O_3, TiC, and Al_2O_3 [19–23]. Among these ceramics, Y_2O_3 was selected as the reinforcement to be used in this study due to its high strength, hardness, melting point, and thermal conductivity [24–26]. Yttria is an air-stable particle, white in color and solid in substance. By adding the yttria to the aluminum, the strength, corrosion resistance, and wear properties are improved [27]. Yttria is well sintered to a high density and low coefficient of thermal expansion, and has excellent strength properties [28,29]. According to the authors' knowledge, there are no reports in the literature on Al–Y_2O_3 nanocomposites processed by mechanical alloying and microwave sintering.

Therefore, in this current research, Al–Y_2O_3 nanocomposites were prepared by mechanical alloying and microwave heating, and the effect of Y_2O_3 addition on the microstructure and mechanical performance of Al–Y_2O_3 nanocomposites were investigated.

2. Materials and Methods

Pure Al (99.5% purity, with an average particle size of 10 μm) and Y_2O_3 nanoparticles (99.99% purity, with an average particle size of 50–70 nm) were purchased from Alfa Aesar (Tewksbury, MA, USA) and selected as raw materials for the synthesis of Al–Y_2O_3 nanocomposites.

Aluminum–yttria composites were prepared with 0, 0.5, 1.0, 1.5, and 2.0 vol. % yttria nanoparticle contents. The mixture of powders was blended at room temperature using a Planetary Ball Mill (PM 200) for 2 h, with a rotation speed of 200 rpm. No balls were used during the blending of powders. The mixed powder (~1.0 gm) was compacted into cylindrical pellets by applying a pressure of 50 MPa with a holding time of 1 min. The compacted cylindrical pellets were sintered in a microwave sintering furnace at a temperature of 550 °C with a heating rate of 10 °C/min and providing a dwell time of 30 min. The microwave furnace had an alumina insulation and silicon carbide susceptor. The silicon carbide susceptor was used to increase the heating rate and hybrid heating. Alumina insulation prevents heat loss and is used as well to protect the interior walls of the microwave oven. The compacted pellets were placed at the center of the cavity and sintering was conducted at the multimode cavity [30]. Figure 1 shows the schematic representation of the microwave sintering furnace.

The density of the sintered samples was calculated using Archimedes' principle. The porosity of the samples was calculated by the theoretical and experimental density of the composite samples. The X-ray diffraction (XRD, PANalytical X'pert Pro, PANalytical B.V., Almelo, The Netherlands) analysis was performed to identify the phases present in Al–Y_2O_3 nanocomposites. The XRD patterns were recorded in the 2θ range of 20–90° with a step size of 0.02° and a scanning rate of 1.5°/min. The microstructural characterization and determination of the distribution of the yttria nanoparticles in the aluminum matrix were carried out using scanning electron microscopy (SEM, JeolNeoscope JSM6000, Tokyo, Japan) and energy dispersive X-ray spectroscopy (EDS, Tokyo, Japan).

The microhardness of the Al–Y_2O_3 nanocomposites was determined using Vickers microhardness tester (MKV-h21, USA). Microhardness analysis was carried out to investigate the effect of yttria on the hardness of the Al–Y_2O_3 nanocomposite, carrying the load of 25 gf and a dwell time of 10 s, for each sample with an average of five successive indentations. Compressive strength analysis was performed at room temperature using a universal testing machine (Lloyd), under an engineering strain rate of 10^{-4}/s.

The respective data of each sample were obtained by an average of three successive values of test results. From the load–displacement curves, 0.2% offset compressive yield strength (CYS), ultimate compressive strength (UCS), and compressive strain were determined.

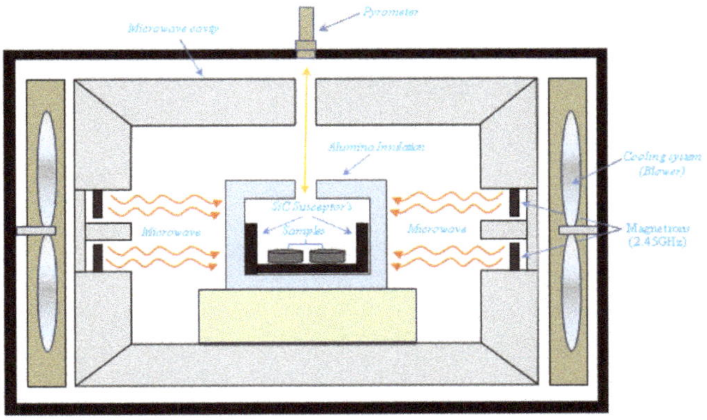

Figure 1. Schematic diagram of a microwave sintering furnace.

3. Results and Discussion

3.1. Density and Porosity of Al–Y_2O_3 Nanocomposites

Density and porosity values of the microwave sintered Al–Y_2O_3 nanocomposites with different contents of yttria in the Al matrix are shown in Table 1.

Table 1. Density and porosity of Al–Y_2O_3 nanocomposites.

Composition	Theoretical Density (g/cc)	Experimental Density (g/cc)	Porosity (%)
Pure Al	2.700	2.679 ± 0.005	0.78
Al–0.5 vol. % Y_2O_3	2.712	2.701 ± 0.004	0.41
Al–1.0 vol. % Y_2O_3	2.723	2.741 ± 0.007	0.33
Al–1.5 vol. % Y_2O_3	2.735	2.728 ± 0.006	0.26
Al–2.0 vol. % Y_2O_3	2.746	2.741 ± 0.008	0.18

It can be observed that the density of the composite gradually increased with the increase of the yttria content since the density of yttria (5.01 g\cc) is higher than that of Al (2.70 g\cc). Generally, the higher relative density of sintered samples influences the mechanical properties of the composites. The porosity of the composites decreased by increasing the amount of yttria content. The decrease in porosity with increasing yttria content shows that the presence of the hard yttria particles did not impair the densification of the Al powder [31]. Microwave heating was one of the main reasons for the low porosity of the synthesized composites.

3.2. XRD Analysis of Al–Y_2O_3 Nanocomposites

The X-ray diffraction (XRD) patterns of the microwave sintered pure Al and Al–Y_2O_3 nanocomposites with different amounts of Y_2O_3 are shown in Figure 2a. Figure 2b shows the enlarged patterns of the Al–1.5 vol. % Y_2O_3 nanocomposite. The XRD patterns clearly indicate the presence of Y_2O_3 nanoparticles in the Al composite matrix. Due to the small volume of yttria reinforcement present in these composites, the yttria peaks were very small compared to the aluminum matrix peaks. Also, it can be seen that the intensity of the yttria diffraction peaks increased with the increasing of yttria percentage. The XRD results show that the main elements of Al (higher peak) and Y_2O_3 (lower peak) are present in Al–Y_2O_3 nanocomposites.

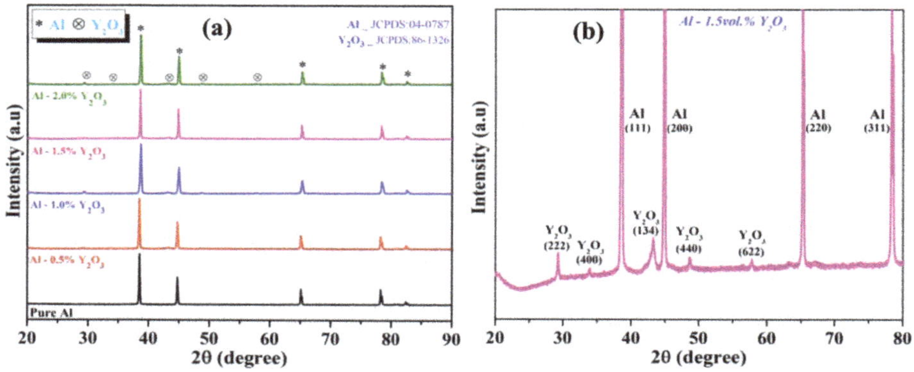

Figure 2. (**a**) X-ray diffraction (XRD) pattern of Al–Y_2O_3 nanocomposites, (**b**) enlarged pattern of Al–1.5vol. %Y_2O_3 nanocomposites [32,33].

3.3. SEM Analysis of Al–Y_2O_3 Nanocomposites

The SEM and EDS images of the microwave sintered Al–Y_2O_3 nanocomposites with different contents of yttria are shown in Figure 3. The results of microstructural characterization revealed that yttria particulates were present individually and in relatively smaller clusters indicating an improvement in their distribution. The EDS analysis confirms the aluminum and yttria particles present in the Al matrix. The EDS mapping spectrum of all nanocomposites were mainly composed of Al, Y, and O elements, as shown in Figure 3b,d,f. The microcracks were restricted by the presence of hard and homogeneous yttria particles in the Al-matrix and influenced the microstructure and mechanical properties of Al–Y_2O_3 nanocomposites. The specimen with 2 vol. % of yttria particles shows the decreasing of the interparticle distances as the concentration of the nanoparticles increased.

3.4. Microhardness of Al–Y_2O_3 Nanocomposites

Vickers microhardness was measured on all specimens to study the effect of Y_2O_3 content on the microhardness. Figure 4 shows the results of the microhardness of the Al–Y_2O_3 nanocomposites with different content of yttria. From the Table 2, the microhardness of the composite increased as the yttria increased of up to 1.5 vol. % and then decreased at 2.0 vol. % Y_2O_3. The considerable increase in hardness could be attributed to the presence of homogeneously distributed hard ceramic nanoparticles and dispersion hardening effect [34]. Al–2.0 vol. % Y_2O_3 nanocomposites show a decreased microhardness value, which was mainly due to the agglomeration of the yttria and increasing presence of clustering of yttria in the case of the Al matrix [35]. The microhardness of the microwave sintered samples in this study was found to be higher than the vacuum sintering and arc-melting samples [36].

The increment of microhardness in the composite materials was due to the presence of hard ceramic particles.

Table 2. Microhardness, yield strength, and ultimate compressive strength of Al–Y_2O_3 nanocomposites.

Composition	Microhardness (Hv)	YS (MPa)	UCS (MPa)	Compression Strain (%)
Pure Al	38 ± 3	69 ± 2	318 ± 5	>60
Al–0.5 vol. % Y_2O_3	46 ± 4	71 ± 4	329 ± 6	>60
Al–1.0 vol. % Y_2O_3	63 ± 2	87 ± 3	337 ± 3	>60
Al–1.5 vol. % Y_2O_3	81 ± 3	126 ± 5	374 ± 6	>60
Al–2.0 vol. % Y_2O_3	74 ± 5	111 ± 5	359 ± 7	>60

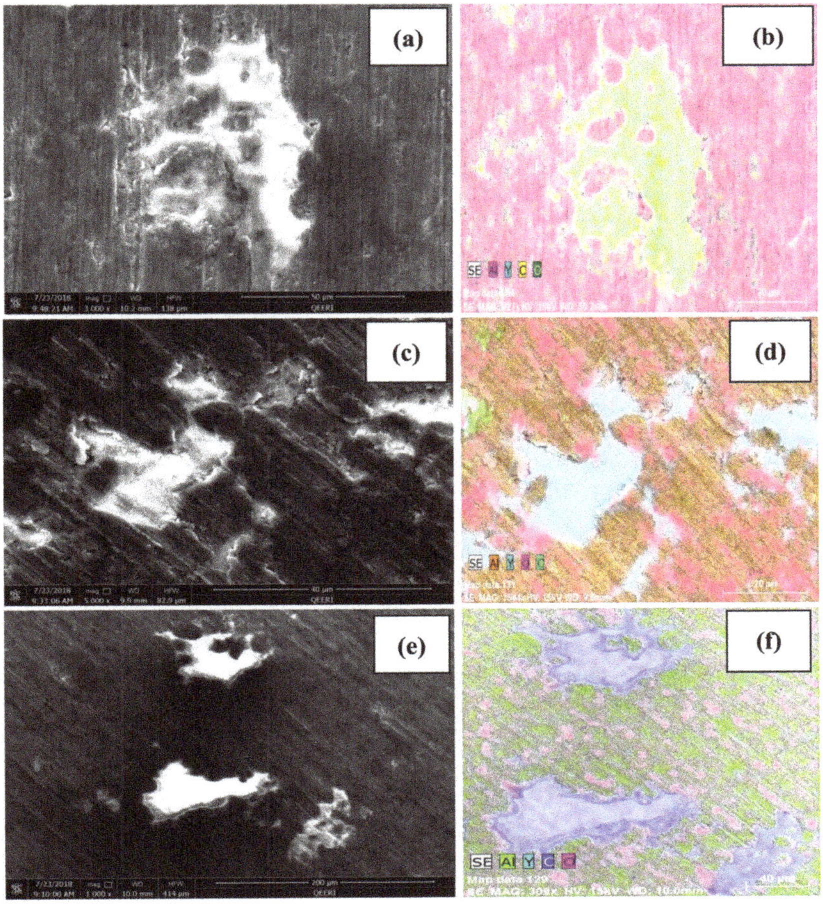

Figure 3. Typical micrographs and corresponding energy dispersion elemental mapping analysis of (**a**–**f**) Al–Y_2O_3 (1, 1.5, and 2 vol. %) nanocomposites.

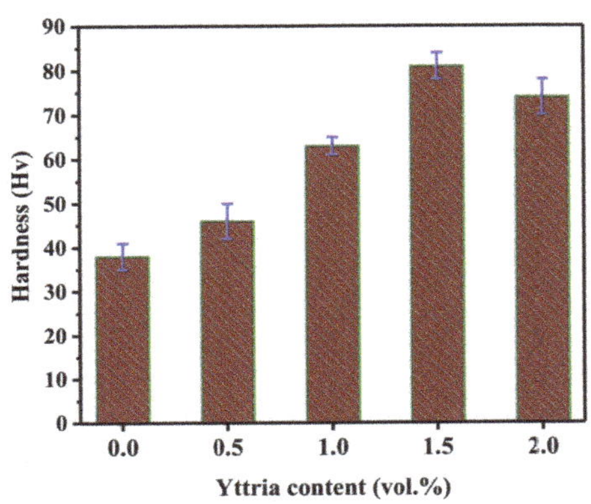

Figure 4. Microhardness of Al–Y_2O_3 nanocomposites.

3.5. Compressive Analysis of Al–Y_2O_3 Nanocomposites

The compressive test was conducted on the microwave sintered pure Al and Al–Y_2O_3 nanocomposites and strengths were compared. Figure 5a shows the engineering stress–strain curves of the Al–Y_2O_3 nanocomposites with different content of yttria. Figure 5b shows the corresponding mechanical data of Al–Y_2O_3 nanocomposites.

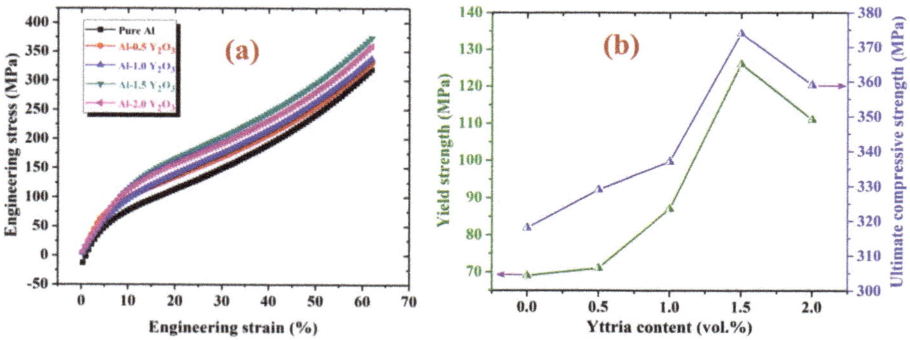

Figure 5. (**a**) The compressive stress–strain curves and (**b**) strength (yield and ultimate) of the Al–Y_2O_3 nanocomposites.

The yield strength and ultimate compressive strength of Al–Y_2O_3 nanocomposites show increased values up to 1.5 vol. % of yttria then decreased as shown in Table 2. Al–1.5 vol. %Y_2O_3 nanocomposites show the maximum yield strength (YS) of 126 ± 5 MPa and ultimate compressive strength (UCS) of 374 ± 6 MPa at a uniform strain of ~60%. These results show the improvement of mechanical properties of Al–Y_2O_3 nanocomposites compared to the pure Al. The increased mechanical properties of the Al–Y_2O_3 nanocomposites are attributed to the dispersion hardening effect and homogeneous distribution of hard reinforcements in the Al-matrix [37]. Al–2.0 vol. % Y_2O_3 nanocomposites show a decreased microhardness value, mainly due to the agglomeration of nanoparticles and grain growth [38]. Reinforcement amounts, density, heating mechanisms factors also govern the variation of the mechanical properties. However, compression properties of the microwave sintered Al–1.5 vol. % Y_2O_3 nanocomposites are interestingly superior to those of other reinforced AMMCs [39–43].

There are several strengthening mechanisms to enhance materials' mechanical properties like hardness and compressive strength of the composite materials. The strengthening of the composites is not only dependent on unique strengthening mechanisms, but it also depends on several strengthening mechanisms.

In the present study, the strengthening mechanism of the Al–Y_2O_3 nanocomposites mainly depended on dispersion hardening due to the hard yttria particles present in the aluminum matrix. The increase in strength and hardness may be attributable to Orowan strengthening [44,45].

3.6. Fractography of Al–Y_2O_3 Nanocomposites

Figure 6 shows the fracture surface images of microwave sintered pure Al and Al–Y_2O_3 nanocomposites under compressive loading. The SEM observations in nanocomposites show typical shear mode fractures and cracks obtained at a 45° to the fracture surfaces with respect to the compressive loading axis. It can be observed that the compressive deformations obtained in pure aluminum and aluminum composites with yttria are different, due to the work hardening behavior. The plastic deformations are restricted by the presence of the second phase in Al–Y_2O_3 nanocomposites [46].

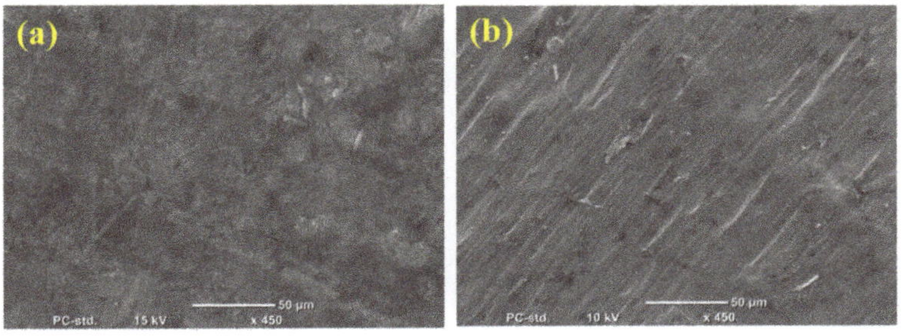

Figure 6. Compression fracture surfaces of (**a**) pure Al and (**b**) Al–1.5 vol. % Y_2O_3 nanocomposites.

4. Conclusions

The Al–Y_2O_3 nanocomposites were successfully synthesized by mechanical alloying and microwave sintering method. The influence of yttria nanoparticles on the microstructure and mechanical properties of the Al–Y_2O_3 nanocomposites were investigated in detail. The density of the composites increased with the increasing of yttria content while porosity decreased. The SEM analysis showed the homogeneous distribution of yttria particles in aluminum composites. The Al–Y_2O_3 nanocomposites exhibited better mechanical properties compared to pure Al. The optimum hardness (81 ± 3 Hv), yield strength (126 ± 5 MPa), and ultimate compression strength (374 ± 6 MPa) and compressive strain (~60%) values were obtained for the Al–1.5 vol. % Y_2O_3 nanocomposite. This significant enhancement in mechanical properties in Al–1.5 vol. % Y_2O_3 nanocomposites make them potential candidates for automotive applications.

Author Contributions: A.S. and A.M.A.M. proposed the original project and supervised the investigation. M.R.M. and P.R.M. performed the experiments, analyzed the data, and wrote the paper with assistance from all authors. All authors contributed to the discussions in the manuscript.

Funding: This publication was made possible by NPRP Grant 7-159-2-076 from the Qatar National Research Fund (a member of the Qatar Foundation). The Qatar National Library funded the publication cost of this article.

Conflicts of Interest: The authors declare no conflict of interest.

References

1. Warren, H.H. Metal matrix composites. In *Comprehensive Composite Materials*; Anthony, K., Carl, Z., Eds.; Pergamon Press: Oxford, UK, 2000; Volume 6, pp. 57–66.
2. Miracle, D.B. Metal matrix composites-from science to technological significance. *Compos. Sci. Technol.* **2005**, *65*, 2526–2540. [CrossRef]
3. Guild, F.J.; Taylor, A.C.; Downes, J. Composite materials. In *Encyclopedia of Maritime and Offshore Engineering*, 1st ed.; John Wiley & Sons, Inc.: Hoboken, NJ, USA, 2017.
4. Zweben, C. Composite materials. In *Mechanical Engineers' Handbook*; Kutz, M., Ed.; Wiley: Hoboken, NJ, USA, 2015.
5. Ghasali, E.; Hossen, A.; Agheli, M. WC-Co particles reinforced aluminum matrix by conventional and microwave sintering. *Mater. Res.* **2015**, *18*, 1197–1202. [CrossRef]
6. Sheasby, P.G.; Pinner, R. *The Surface Treatment and Finishing of Aluminum and Its Alloys*, 6th ed.; Finishing Publications Ltd.: Stevenage, UK; ASM International: Novelty, OH, USA, 2001; pp. 1–596.
7. Pardeep, S.; Satpal, S.; Dinesh, K. A study on the microstructure of aluminum matrix composites. *J. Asian Ceram. Soc.* **2105**, *3*, 240–244.
8. Manoj, S.; Deepak, D.D.; Lakhvir, S.; Vikas, C. Development of aluminium based silicon carbide particulate metal matrix composite. *J. Miner. Mater. Charact. Eng.* **2009**, *8*, 455–467.
9. Rana, R.S.; Purohit, R. Reviews on the influences of alloying elements on the microstructure and mechanical properties of aluminum alloys and aluminum alloy composites. *J. Sci. Eng. Res.* **2012**, *2*, 1–7.

10. Bharath, V.; Madev, N.; Auradi, V.; Kori, S.A. Preparation of 6061Al-Al_2O_3 MMC's by stir casting and evaluation of mechanical and wear properties. *Procedia Mater. Sci.* **2014**, *6*, 1658–1667. [CrossRef]
11. Narayan, S.; Rajeshkannan, A. Hardness, tensile and impact behavior of hot forged aluminum metal matrix composites. *J. Mater. Res. Technol.* **2017**, *6*, 213–219. [CrossRef]
12. Ghasali, E.; Pakseresht, A.; Ebadzadeh, T. Investigation on microstructure and mechanical behavior of Al–ZrB_2 composite prepared by microwave and spark plasma sintering. *Mater. Sci. Eng. A* **2015**, *627*, 27–31. [CrossRef]
13. Kong, J.; Xu, C.; Li, J.; Hou, H. Evolution of fractal features of pores in the compacting and sintering process. *Adv. Powder Technol.* **2011**, *22*, 439–442. [CrossRef]
14. Breval, E.; Cheng, J.P.; Agrawal, D.K.; Gigl, P.; Dennis, M.; Roy, R.; Papworth, A.J. Comparison between microwave and conventional sintering of WC/Co composites. *Mater. Sci. Eng. A* **2005**, *391*, 285–295. [CrossRef]
15. Shen, Z.; Johnsson, M.; Zhao, Z.; Nygren, M. Spark plasma sintering of alumina. *J. Mater. Process. Technol.* **2002**, *85*, 1921–1927. [CrossRef]
16. Jin, L.; Zhou, G.; Shimai, S.; Zhang, J.; Wang, S. ZrO_2-doped Y_2O_3 transparent ceramics via slip casting and vacuum sintering. *J. Eur. Ceram. Soc.* **2010**, *30*, 2139–2143. [CrossRef]
17. Rajkumar, K.; Aravindan, S. Microwave sintering of copper–graphite composites. *J. Mater. Process. Technol.* **2009**, *209*, 5601–5605. [CrossRef]
18. Reddy, M.P.; Ubaid, F.; Shakoor, R.A.; Gururaj, P.; Vyasaraj, M.; Mohamed, A.M.A.; Gupta, M. Effect of reinforcement concentration on the properties of hot extruded Al-Al_2O_3 composites synthesized through microwave sintering process. *Mater. Sci. Eng. A* **2017**, *696*, 60–69. [CrossRef]
19. Reddy, M.P.; Ubaid, F.; Shakoor, R.A.; Gururaj, P.; Vyasaraj, M.; Mohamed, A.M.A.; Gupta, M. Enhanced performance of nano-sized SiC reinforced Al metal matrix nanocomposites synthesized through microwave sintering and hot extrusion techniques. *Prog. Nat. Sci. Mater.* **2017**, *27*, 607–614. [CrossRef]
20. Wang, M.; Wang, D.; Kups, T.; Schaaf, P. Size effect on mechanical behavior of Al/Si_3N_4 multilayers by nanoindentation. *Mater. Sci. Eng. A* **2015**, *644*, 275–283. [CrossRef]
21. Kim, C.S.; Kim, I.H. Effect of Al and Y_2O_3 on mechanical properties in mechanically alloyed nanograin Ni-based alloys. *J. Nanosci. Nanotechnol.* **2105**, *15*, 6160–6163. [CrossRef]
22. Karantzalis, A.E.; Wyatt, S.; Kennedy, A.R. The mechanical properties of Al-TiC metal matrix composites fabricated by a flux-casting technique. *Mater. Sci. Eng. A* **1997**, *237*, 200–206. [CrossRef]
23. Hossein-Zadeh, M.; Razavi, M.; Mirzaee, O.; Ghaderi, R. Characterization of properties of Al–Al_2O_3 nano-composite synthesized via milling and subsequent casting. *J. King Saud Univ. Eng. Sci.* **2013**, *25*, 75–80. [CrossRef]
24. Ianos, R.; Babuta, R.; Lazau, R. Characteristics of Y_2O_3 powders prepared by solution combustion synthesis in the light of a new thermodynamic approach. *Ceram. Int.* **2014**, *40*, 12207–12211. [CrossRef]
25. Chaim, R.; Shlayer, A.; Estournes, C. Densification of nanocrystalline Y_2O_3 ceramic powder by spark plasma sintering. *J. Eur. Ceram. Soc.* **2009**, *29*, 91–98. [CrossRef]
26. Kakuoz, Y.B.; Serivalsatit, K.; Kokuoz, B.; Geiculescu, O.; McCormick, E.; Ballato, J. Er-doped Y_2O_3 nanoparticles: A comparison of different synthesis methods. *J. Am. Ceram. Soc.* **2009**, *92*, 2247–2253. [CrossRef]
27. Bouaeshi, W.B.; Li, D. Effects of Y_2O_3 addition on microstructure, mechanical properties, electrochemical behavior, and resistance to corrosive wear of aluminum. *Tribol. Int.* **2007**, *40*, 188–199. [CrossRef]
28. Cho, J.; Harmer, M.P.; Chan, H.M.; Rickman, J.M.; Thompson, A.M. Effect of yttrium and lanthanum on the tensile creep behavior of aluminum oxide. *J. Am. Ceram. Soc.* **1993**, *28*, 6466–6473. [CrossRef]
29. Borovkova, L.B.; Lukin, E.S.; Poluboyarinov, D.N.; Sapozhnikova, E.B. Sintering and properties of yttrium oxide ceramics. *Refract. Ind. Ceram.* **1970**, *11*, 717–722. [CrossRef]
30. Ranjan, S. Sintering and Mechanical Properties of Alumina-Yttrium Aluminate Composites. Ph.D. Thesis, National Institute of Technology, Rourkela, India, May 2015.
31. Meena, K.L.; Manna, A.; Banwait, S.S.; Jaswanti, D. An analysis of mechanical properties of the developed Al/SiC-MMC's. *Am. J. Mech. Eng.* **2013**, *1*, 14–19. [CrossRef]
32. Vamsi Krishna, M.; Xavior, M.A. Experiment and statistical analysis of end milling parameters for al/sic using response surface methodology. *Int. J. Eng. Technol.* **2015**, *7*, 2274–2285.

33. Dong, G.; Chi, Y.; Xiao, X.; Liu, X.; Qian, B.; Ma, Z.; Wu, E.; Zeng, H.; Chen, D.; Qiu, J. Fabrication and optical properties of Y_2O_3: Eu^{3+} nanofibers prepared by electrospinning. *Opt. Express* **2009**, *17*, 22514–22519. [CrossRef]
34. Reddy, M.P.; Himyan, M.A.; Ubaid, F.; Shakoor, R.A.; Gururaj, P.; Vyasaraj, M.; Mohamed, A.M.A.; Gupta, M. Enhancing thermal and mechanical response of aluminium using nanolength scale TiC reinforcement. *Ceram. Int.* **2018**, *44*, 9247–9254. [CrossRef]
35. Hassan, S.F.; Gupta, M. Development of nano-Y_2O_3 containing magnesium nanocomposites using solidification processing. *J. Alloy Compd.* **2007**, *429*, 176–183. [CrossRef]
36. Zhao, N.Q.; Jiang, B.; Du, X.W.; Li, J.J.; Shi, C.S.; Zhao, W.X. Effect of Y_2O_3 on the mechanical properties of open cell aluminum foams. *Mater. Lett.* **2006**, *60*, 1665–1668. [CrossRef]
37. Reddy, M.P.; Ubaid, F.; Shakoor, R.A.; Parande, G.; Vyasaraj, M.; Yusuf, M.; Mohamed, A.M.A.; Gupta, M. A comparative study of structural and mechanical properties of Al-Cu composites prepared by vacuum and microwave sintering techniques. *J. Mater. Res. Technol.* **2017**, *7*, 165–172.
38. Pasquale, C.; Farhad, J.; Ali, S.; Behzad, S. Influence of SiO_2 nanoparticles on the microstructure and mechanical properties of Al matrix nanocomposites fabricated by spark plasma sintering. *Compos. Part B Eng.* **2108**, *146*, 60–68.
39. Mahmut, C.S.; Gürbüz, M.; Koç, E. Fabrication and characterization of SiC and Si_3N_4 reinforced aluminum matrix composites. *Univers. J. Mater. Sci.* **2017**, *5*, 95–101.
40. Pakdel, A.; Farhangi, H.E. Effect of extrusion process on ductility and fracture behavior of SiCp/aluminum alloy composites. Proceedings of 8th International Fracture Conference, Istanbul Turkey, 7–9 November 2007.
41. Moustafa, M.M.M.; Omayma, A.E.; Abdelhameed, W.A. Effect of alumina particles addition on physico-mechanical properties of Al-matrix composites. *Open J. Metal* **2013**, *3*, 72–79.
42. Ehsan, G.; Masoud, A.; Touradj, E.; Amir, H.P.; Ali, R. Investigation on microstructural and mechanical properties of B4C–aluminum matrix composites prepared by microwave sintering. *J. Mater. Res. Technol.* **2015**, *4*, 411–415.
43. Hasan, K.I.; Aboozar, T.; Ali, M.; Abbas, G. Development of an aluminum/amorphous nano-SiO_2 composite using powder metallurgy and hot extrusion processes. *Ceram. Int.* **2017**, *43*, 14582–14592.
44. Zeng, X.; Liu, W.; Xu, B.; Shu, G.; Li, Q. Microstructure and mechanical properties of Al-SiC nanocomposites synthesized by surface-modified aluminium powder. *Metals* **2018**, *8*, 253. [CrossRef]
45. Ramezanalizadeh, H.; Emamy, M.; Shokouhimehr, M. A novel aluminum based nanocomposite with high strength and good ductility. *J. Alloys Compd.* **2015**, *649*, 461–473. [CrossRef]
46. Reddy, M.P.; Ubaid, F.; Shakoor, R.A.; Parande, G.; Vyasaraj, M.; Yusuf, M.; Mohamed, A.M.A.; Gupta, M. Improved properties of Al–Si_3N_4 nanocomposites fabricated through a microwave sintering and hot extrusion process. *RSC Adv.* **2017**, *7*, 34401–34410.

metals

MDPI

Article

Mechanical Amorphization and Recrystallization of Mn-Co(Fe)-Ge(Si) Compositions

Antonio Vidal-Crespo, Jhon J. Ipus, Javier S. Blázquez * and Alejandro Conde

Departamento Física de la Materia Condensada, ICMSE-CSIC, Universidad de Sevilla, 41080 Sevilla, Spain; antvidcre@alum.us.es (A.V.-C.); jhonipus@us.es (J.J.I.); conde@us.es (A.C.)

* Correspondence: jsebas@us.es; Tel.: +34-954-556-029

Received: 18 April 2019; Accepted: 6 May 2019; Published: 8 May 2019

Abstract: Mechanical alloying using a planetary ball mill allowed us to obtain two homogeneous systems formed by units with nanometer size and $MnCo_{0.8}Fe_{0.2}Ge_{1-x}Si_x$ stoichiometry ($x = 0$ and 0.5). The phase evolution of the systems with the milling time was analyzed using X-ray diffraction. Thermal stability of the final products was studied using differential scanning calorimetry. Room temperature ^{57}Fe Mössbauer spectroscopy was used to follow the changes in the Fe environments. A paramagnetic Co-based amorphous phase developed in both alloys as milling progressed. However, while the presence of Si stabilized the Mn-type phase, mechanical recrystallization was observed in a Si-free composition leading to the formation of a MnCo(Fe)Ge intermetallic (*Pnma* space group) with a crystal size of 7 ± 1 nm. Mössbauer results indicate that Fe atoms migrate from the initial bcc phase to the amorphous and intermetallic phases.

Keywords: half-Heusler alloys; mechanical alloying; Mössbauer spectroscopy

1. Introduction

Half-Heusler MnCoGe alloys can show a martensitic transformation from an orthorhombic TiNiSi-type structure (*Pnma* space group) to a hexagonal Ni_2In-type structure ($P6_3/mmc$, although it can be also interpreted as a different orthorrombic *Pnma* structure with different lattice parameters to those of the TiNiSi-type structure [1]). Coincidence of such a structural transformation with a magnetic one has been proposed to enhance the magnetocaloric effect exhibited by these systems [2], which can be achieved by compositional tailoring with partial substitution of Fe for Co [3]. However, the formation of the intermetallic phase of interest is not straightforward and long-duration annealing at high temperatures is needed (typically several days at ~1125 K [3–5]).

On the other hand, high entropy alloys (HEAs) are homogeneous solid solutions formed by at least five different elements with atomic fractions between 5 and 35 at. % [6]. In such HEAs, bcc and fcc solid solutions as well as amorphous phases can be observed as product phases when produced by rapid quenching [7] or mechanical alloying [8]. Both amorphous and supersaturated solid solutions are very attractive homogeneous precursor systems to develop stoichiometric intermetallic phases, strongly reducing the annealing time required with respect to the samples obtained by conventional methods [9]. The development of HEAs from half-Heusler compositions has been recently reported for Ti(NiCo)(SnSb) [10], CoMnSn(Cu) [11] and (TiZr)Ni(SnSb) [12] systems.

The aim of this study is to produce homogeneous systems starting from pure powders with $MnCo_{0.8}Fe_{0.2}Ge_{1-x}Si_x$ stoichiometry (with $x = 0$ and $x = 0.5$) using mechanical alloying.

2. Materials and Methods

Pure Mn (99.6%, Alfa Aesar, Karlsruhe, Germany), Co (99.99%, Chempur, Karlsruhe, Germany), Ge (99.99%, Chempur, Karlsruhe, Germany), Si (99.9%, Alfa Aesar, Karlsruhe, Germany) and Fe (>99%,

Alfa Aesar, Karlsruhe, Germany) powders (5 g) were mixed in hardened steel vials with 10 mm steel balls in an argon atmosphere and ball milled up to 100 h at 250 rpm in a Pulverisette Vario 4 mill (Fritsch, Idar-Oberstein, Germany) with a frequency ratio of −2 and a ball mass to powder ratio, BPR = 10. Compositions were checked using EAGLE III (EDAX, Mahwah, NJ, USA) X-ray microfluorescence equipment. X-ray diffraction (XRD) experiments were performed using a powder diffractometer D8 Advance A25 (Bruker, Karlsruhe, Germany) at room temperature and the radiation employed was Cu K_{α}. Experimental patterns were fitted using TOPAS software (Version 6, Bruker, Karlsruhe, Germany). No preferential orientation was allowed to preserve the intensity ratio in our powder samples. Transmission ^{57}Fe Mössbauer (MS) spectra at room temperature were obtained using a Wissel spectrometer (Wissel, Starnberg, Germany). Isomer shifts were measured relative to that of a standard foil of pure Fe. Differential scanning calorimetry (DSC) experiments were performed using a DSC7 (Perkin-Elmer, Norwalk, CT, USA) calorimeter at a heating rate of 20 K/min.

3. Results

3.1. X-ray Diffraction

Figure 1 shows the evolution of the XRD patterns as a function of the milling time. These patterns can be fitted using the Rietveld method assuming the different starting pure phases, except for Si (the lightest among the studied elements), which is not detected even after 1 h milling, indicating its integration to the other phases. Table 1 shows the R-factors of the different fittings and Table 2 shows the lattice parameter of the different phases detected. The diffraction maxima of the starting hcp Co phase rapidly broadens beyond any realistic values of crystal size or microstrains, which is due to the formation of an amorphous phase in both compositions. In order to account for this amorphous phase, we allowed the amorphous halo associated to this phase to evolve directly from the diffraction maxima of the hcp Co phase. Although Rietveld fitting of an amorphous phase could lead to unphysical results of the parameters (e.g., extremely low crystal size or extremely high microstrains), our aim was just to estimate the phase fraction evolution along the milling.

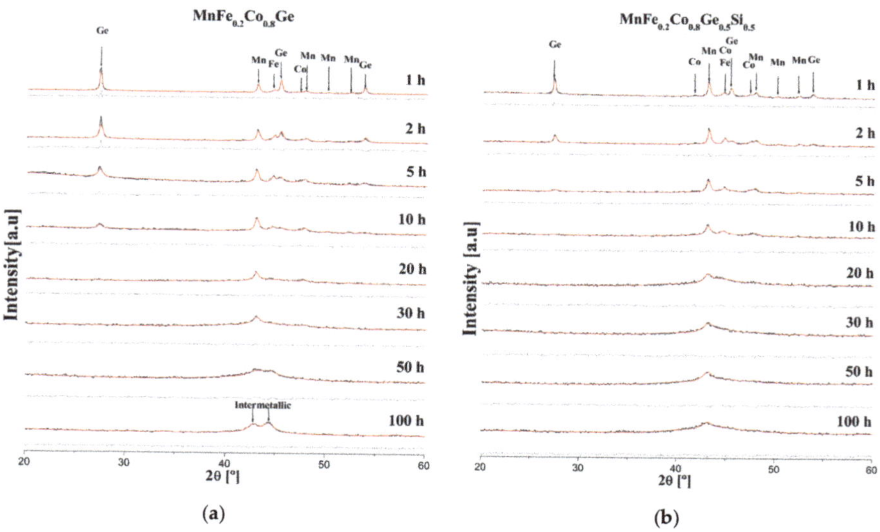

Figure 1. X-ray diffraction (XRD) patterns of samples after different times of milling: (**a**) Si-free alloy (**b**) Si-containing alloy. The corresponding differences between the experimental data and the Rietveld fittings are shown below each experimental pattern. The experimental data in black and the fitting in red.

Rietveld fitting (see R-values in Table 1 for each pattern) supplies valuable information concerning phase fraction, lattice parameter, crystal size and microstrains. In the following we will account only for the crystal size as the main factor for peak broadening (i.e., a minimum crystal size is reported). Figure 2 shows the phase fraction evolution with the milling time of the two studied compositions and Figure 3 shows the corresponding crystal size.

Table 1. Parameters from Rietveld fittings.

Si-Free Alloy				Si-Containing Alloy			
Milling Time (h)	R_{exp}	R_{wp}	χ^2	**Milling Time (h)**	R_{exp}	R_{wp}	χ^2
1	2.20	3.87	1.76	1	2.36	2.90	1.23
2	2.23	3.37	1.51	2	2.33	2.61	1.12
5	2.29	2.98	1.30	5	2.17	2.44	1.12
10	2.24	2.60	1.16	10	2.26	2.39	1.06
20	2.01	2.07	1.03	20	2.21	2.59	1.17
30	2.01	2.08	1.07	30	2.21	2.54	1.15
50	2.02	2.36	1.17	50	2.22	2.31	1.04
100	2.07	2.28	1.10	100	2.08	2.28	1.10
Crystalline	2.16	3.08	1.43	Crystalline	2.22	3.37	1.52

Table 2. Average lattice parameters of the crystalline phases detected by XRD. Changes in this parameter with the milling time is of the order of the error bar.

Phase	**Space Group**	**Lattice Parameter (Å)**	
		Si-Free	**Si-Containing**
Mn	*I4-3m*	8.916 ± 0.003	8.893 ± 0.009
Ge	*Fd-3m*	5.656 ± 0.002	5.656 ± 0.006
Fe(Co)	*Im-3m*	2.870 ± 0.002	2.870 ± 0.002
MnCo(Fe)Ge	*Pnma*	a = 5.20 ± 0.02 b = 4.15 ± 0.04 c = 7.0 ± 0.2	-
MnCo(Fe)Ge *	*Pnma*	a = 5.2822 ± 0.0002 b = 4.0750 ± 0.0003 c = 7.0440 ± 0.0005	-
Bcc solid solution *	*Im-3m*	-	2.8835 ± 0.0001

* Samples heated up to 973 K at 20 K/min in argon flow.

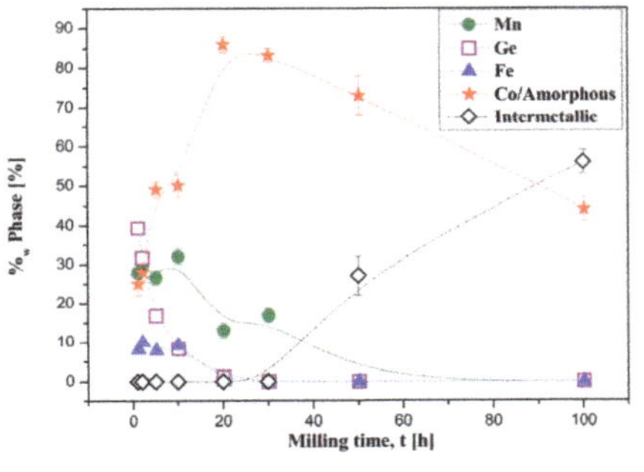

(a)

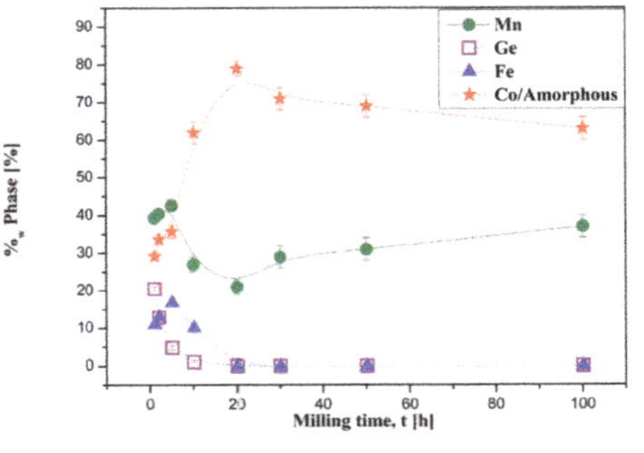

(b)

Figure 2. Phase fraction from XRD Rietveld analysis as a function of the milling time: (**a**) Si-free alloy (**b**) Si-containing alloy. Lines are a guide to the eye.

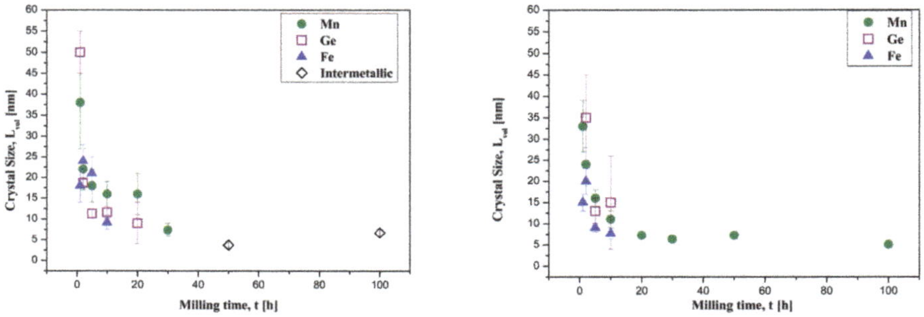

(a) (b)

Figure 3. Crystal size from XRD Rietveld analysis as a function of the milling time: (**a**) Si-free alloy (**b**) Si-containing alloy.

The fraction of the diamond-like Ge phase (*Fd-3m* space group) exponentially decreased with milling time for both studied compositions. The content of bcc-Fe-type phase (*Im-3m* space group) initially reached values above the starting weight fraction of Fe, indicating the migration of other atoms (mainly Co and Si) to this phase. After 10 h of milling, the decrease of the weight fraction for this phase was clear and was no longer detected by XRD after 20 h milling. The evolution of the Mn phase fraction depends on the Si content of the sample. Whereas for Si-free alloy, the Mn-type phase was no longer detected by XRD after 50 h milling; for Si-containing alloy, the Mn-type phase remained almost constant (or even increased) from 10 h up to the maximum time explored in this study (100 h).

As written above, Rietveld fitting showed that the crystal size of the Co-type phase ($P6_3/mmc$ space group) rapidly decreased with milling time (below 2 nm after 5 h milling for Si-free alloy and after 10 h milling for Si-containing alloy). Therefore, the diffraction maxima ascribed to this phase no longer describe a crystalline phase but rather amorphous halos ascribed to an amorphous phase. The distinction between crystalline Co-type and Co-base amorphous phase is not clear. Therefore, the two phases are represented together in Figure 2.

Lattice parameters did not change significantly with the milling time but the average values could differ with respect to the values of the pure phases. This should indicate that e.g., Co migration to bcc Fe occurs at the early stages of milling, in agreement with the higher fraction measured for this phase (8–10 wt. %) with respect to the nominal Fe fraction (6.0 and 6.8 wt. % for the alloys without and with Si). In the case of Ge, the measured lattice parameter agreed with that of the pure phase, indicating that this element preserves its purity during its comminution. In the case of the Mn phase, the presence of Si stabilizes it and reduces the lattice parameter with respect to that of the Si-free composition. Average values of lattice parameters of the different phases are shown in Table 2.

3.2. *Mössbauer Spectroscopy*

Figure 4 shows the evolution of the Mössbauer spectra with the milling time. Two main contributions can be clearly distinguished: a ferromagnetic contribution (FM) and a paramagnetic (PM) one. The FM contribution corresponds to Fe atoms in the bcc Fe(Co) phase as it is confirmed by the hyperfine field HF~33 T. This may indicate that Fe content in crystalline hcp Co at the early stages is negligible (no site is detected). As milling time increased, FM contribution reduced to zero at around 20 h milling, whereas the PM contribution was present since the earlier studied times and progressively increased with milling. Therefore, as the FM contribution is clearly assigned to bcc-Fe type sites, the rest of the phases, which contain Fe, detected by XRD, must be paramagnetic, including the amorphous phase derived from broadening of the hcp-Co diffraction maxima.

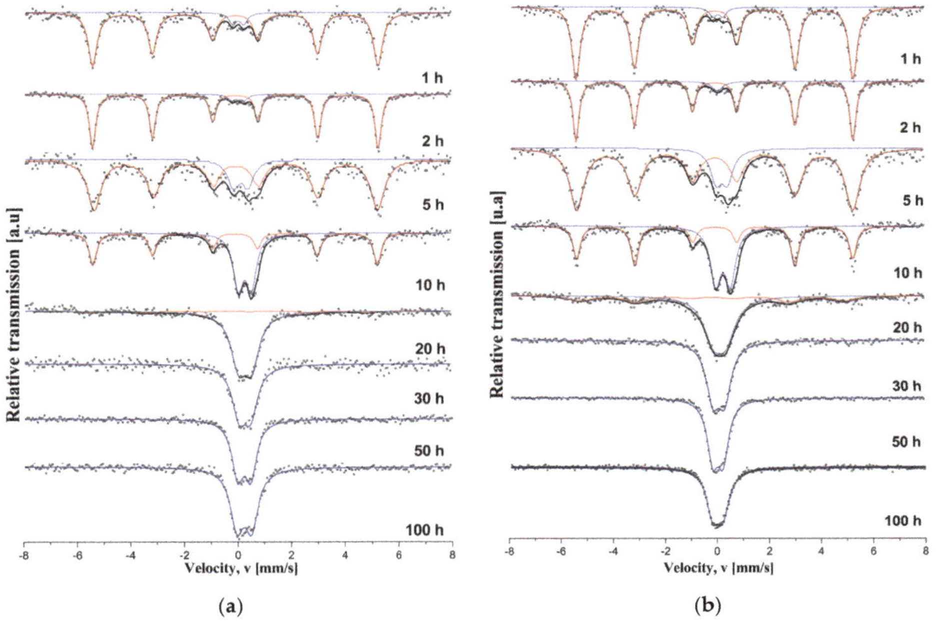

Figure 4. Room temperature Mössbauer spectra as a function of the milling time of: (**a**) Si-free alloy and (**b**) Si-containing alloy samples.

4. Discussion

The recrystallization process was detected only in Si-free alloys, leading to the formation of an intermetallic phase: MnCo(Fe)Ge (*Pnma* space group), with a crystal size <10 nm for as-milled samples after 50 and 100 h (see Figure 3a). The presence of the recrystallization phenomenon was confirmed by DSC. Figure 5a shows the DSC scan of Si-free samples milled for 50 h and 100 h, respectively. The transformation heat, $|\Delta H|$, of the exothermic peak at ~550 K strongly decreased from the sample milled for 50 h to the sample milled for 100 h (from $\Delta H = -38$ to -16 ± 1 J/g, while the amorphous fraction from XRD decreased from 73 to 44%) due to the recrystallization phenomenon. In fact, XRD patterns of samples heated above the exothermic peak showed the intermetallic MnCoGe-type phase as the single phase present except for some traces of MnO (as shown in Figure 6).

In the case of the Si-containing sample after 100 h milling, DSC of Figure 5b shows a minor exothermic peak at ~550 K but the main transformation peak is found at ~620 K ($\Delta H = -114 \pm 1$ J/g). Samples heated above this temperature transformed to a single bcc solid solution. At higher temperatures, ~850 K, an endothermic and reversible peak was found.

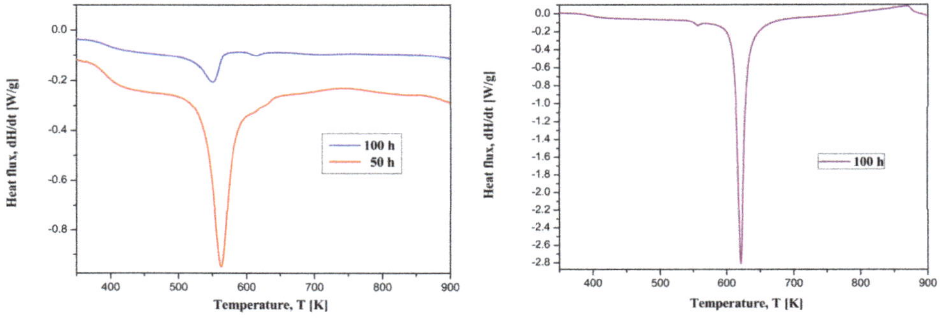

(a) (b)

Figure 5. DSC scans at 20 K/min for: (**a**) Si-free alloy after 50 and 100 h milling (**b**) Si-containing alloy after 100 h milling.

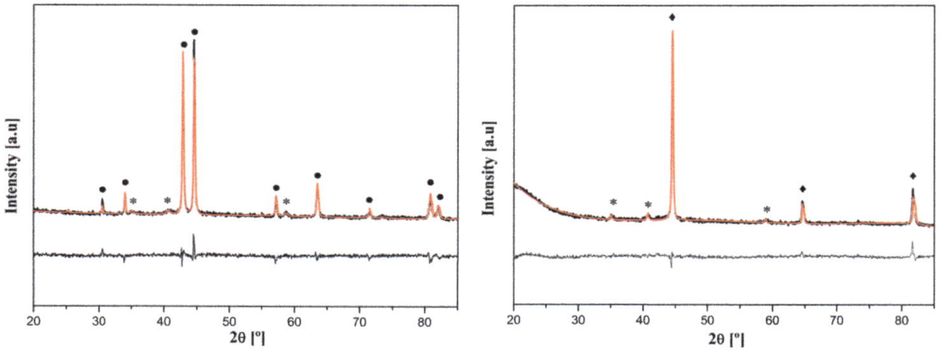

(a) (b)

Figure 6. XRD patterns of samples milled for 100 h and heated up to 973 K at 20 K/min: (**a**) Si-free alloy (**b**) Si-containing alloy. Circles identify intermetallic phase, diamonds identify bcc solid solution and asterisks identify MnO phase. The experimental data are in black and the fitting in red. Corresponding differences between the experimental and fitting curves are shown below each experimental pattern.

The formation of such simple structures in Si-containing alloy (amorphous and bcc solid solution) is typically found in HEA. These systems can be characterized by several parameters:

Mixing enthalpy, ΔH_{mix};

$$\Delta H_{mix} = 4\sum_{i,j<i} c_i c_j \Delta H^{ij}_{mix}, \quad (1)$$

with c_i the molar fraction of the i element in the composition and ΔH^{ij}_{mix}, the mixing enthalpy between the elements i and j. For non-metals such as Si and Ge, it is necessary to calibrate ΔH^{ij}_{mix} to subtract the extra energy, ΔH^{trans}, required to transform the element from non-metallic to metallic and: $\Delta H^{ij*}_{mix} = \Delta H^{ij}_{mix} - \Delta H^{trans}/2$. For Si and Ge, ΔH^{trans} of 34 or 25 kJ/mol [13], respectively.

Atomic-size difference, δ:

$$\delta = \sqrt{\sum_{i=1}^{N} c_i \left(1 - \frac{r_i}{\sum_{i=1}^{N} c_i r_i}\right)^2}, \tag{2}$$

where r_i is the atomic radius of the i element. And Ω parameter:

$$\Omega = \frac{T_m \Delta S_{mix}}{\Delta H_{mix}}, \tag{3}$$

where T_m is the weighted average of the melting temperature of the composition and

$$\Delta S_{mix} = -R \sum_{i} c_i ln(c_i), \tag{4}$$

Taking the values of r_i and ΔH^{ij}_{mix} from [6,13], the results for the quinary composition are ΔH_{mix} = −33 kJ/mol, δ = 5.4% and Ω = 0.60. While the quaternary composition has ΔH_{mix} = −25 kJ/mol, δ = 3.9% and Ω = 0.64. These parameters place our studied alloys close to the bulk amorphous region of HEA depicted in [6]. This could agree with the easy formation of the amorphous phase in our studied samples. However, although the studied quinary composition develops a single bcc phase solid solution after thermal treatment, HEA with such solid solution generally shows larger Ω and less negative ΔH_{mix} [6].

Despite the low content of Fe in the studied compositions (~6.7 at. %), MS can supply some information to confirm the evolution of the phases during milling. As already described, the only FM site detected corresponds to the bcc-Fe(Co) phase and was fitted using a broad sextet centered at HF~33 T, which confirms the rapid transformation of the FM hcp-Co phase to a PM Co-based amorphous phase enriched in Ge and Mn (or at least the negligible Fe content in the residual hcp-Co phase).

The PM contribution was fitted using a doublet with quadrupolar splitting, <Q> = 0.48 ± 0.07 mm/s (Si-free alloy) and 0.42 ± 0.15 mm/s (Si-containing alloy). In the case of Si-free alloy, both *Q* and *IS* remain almost constant along the milling process indicating that the Fe sites out of the bcc phase might be similar (i.e., Fe atoms are not expected to migrate to many different phases). In the case of Si-containing alloy, whereas *Q* is almost constant, *IS* becomes more positive as milling time increases. A clear correlation can be observed between the area fraction of the PM contribution and the amorphous fraction measured from XRD for both compositions up to 30 and 20 h of milling for Si-free and Si-containing alloy, respectively. Figure 7 shows this correlation. In the case of Si-free alloy, correlation is preserved for all times when amorphous and intermetallic phase fractions are considered. This indicates that Fe atoms migrate to these two phases in Si-free alloy.

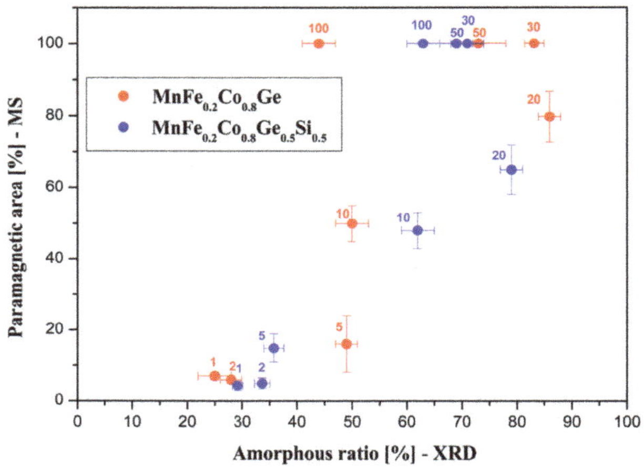

Figure 7. Amorphous fraction from XRD Rietveld analysis as a function of PM contribution from MS for Si-free alloy (in red) and Si-containing alloy (in blue). Numbers indicate the milling time in hours for each composition.

5. Conclusions

Two different compositions $MnCo_{0.8}Fe_{0.2}(Ge_{1-x}Si_x)$ were partially amorphized by mechanical alloying. X-ray diffraction and Mössbauer spectrometry were used to characterize the evolution of the different phases with milling time.

In the case of the Si-free alloy, almost fully amorphization was achieved after 50 h milling, and further milling led to the development of MnCoGe-type intermetallic. Thermal treatment beyond ~650 K led to the growth of this intermetallic, and the alloy became single phase.

In the case of the Si-containing alloy, the Mn-phase fraction remained almost constant from 30 to 100 h milling and the alloy became only partially amorphous during milling. Thermal treatment beyond ~650 K led to the formation of a bcc solid solution, which is characteristic for high entropy alloys.

Author Contributions: Conceptualization of the project: A.V.-C., J.J.I., J.S.B. and A.C. Discussion of the results and revision of the paper: A.V.-C., J.J.I., J.S.B. and A.C. Experiments were developed by A.V.-C., J.J.I., and J.S.B.

Funding: This research was funded by AEI/FEDER-UE (Project MAT 2016-77265-R) and the PAI of the Regional Government of Andalucía.

Conflicts of Interest: The authors declare no conflict of interest.

References

1. Jeitschko, W. A High-Temperature X-ray Study of the Displacive Phase Transition in MnCoGe. *Acta Crystallogr. Sect. B: Struct. Sci.* **1975**, *31*, 1187–1190. [CrossRef]
2. Johnson, V. Diffusionless orthorhombic to hexagonal transitions in ternary silicides and germanides. *Inorg. Chem.* **1975**, *14*, 1117–1120. [CrossRef]
3. Li, G.; Liu, E.; Zhang, H.; Zhang, Y.; Chen, J.; Wang, W.; Zhang, H.; Wu, H.; Yu, H. Phase diagram, ferromagnetic martensitic transformation and magnetoresponsive properties of Fe-doped MnCoGe alloys. *J. Magn. Magn. Mater.* **2013**, *332*, 146–150. [CrossRef]
4. Ozono, K.; Mitsui, Y.; Umetsu, R.; Hiroi, M.; Takahashi, K.; Koyama, K. Magnetic and structural properties of MnCo1-xFexGe ($0 \leq x \leq 0.12$). *AIP Conf. Proc.* **2016**, *1763*. [CrossRef]
5. Lin, S.; Tegus, O.; Bruck, E.; Dagula, W.; Gortenmulder, T.; Buschow, K. Structural and Magnetic Properties of $MnFe_1$-xCoxGe Compounds. *IEEE Trans. Mag.* **2006**, *42*, 3776–3778. [CrossRef]
6. Zhang, Y.; Zuo, T.; Tang, Z.; Gao, M.; Dahmen, K.; Liaw, P.; Lu, Z. Microstructures and properties of high-entropy alloys. *Prog. Mater. Sci.* **2014**, *61*, 1–93. [CrossRef]

7. Singh, S.; Wanderka, N.; Murty, B.; Glatzel, U.; Banhart, J. Decomposition in multi-component AlCoCrCuFeNi high-entropy alloy. *Acta Mater.* **2011**, *59*, 182–190. [CrossRef]
8. Chen, Y.; Tsai, C.; Juan, C.; Chuang, M.; Yeh, J.; Chin, T.; Chen, S. Amorphization of equimolar alloys with HCP elements during mechanical alloying. *J. Alloys Compd.* **2010**, *506*, 210–215. [CrossRef]
9. Blázquez, J.S.; Ipus, J.J.; Moreno-Ramírez, L.M.; Álvarez-Gómez, J.M.; Sánchez-Jiménez, D.; Lozano-Pérez, S.; Franco, V.; Conde, A. Ball milling as a way to produce magnetic and magnetocaloric materials: A review. *J. Mater. Sci.* **2017**, *52*, 11834–11850. [CrossRef]
10. Karati, A.; Nagini, M.; Ghosh, S.; Shabadi, R.; Pradeep, K.G.; Mallik, R.C.; Murty, B.S.; Varadaraju, U.V. $Ti_2NiCoSnSb$—A new halfHeusler type high-entropy alloy showing simultaneous increase in Seebeck coefcient and electrical conductivity for thermoelectric applications. *Sci. Rep.* **2019**, *9*, 5331. [CrossRef]
11. Guo, Z.; Qiu, H.; Liu, Z. Effects of the substitution of Cu for Sn on structural, magnetic andmagnetocaloric properties of half-Heusler CoMnSn alloy. *J. Alloys Compd.* **2019**, *777*, 472–477. [CrossRef]
12. Rogl, G.; Yubuta, K.; Romaka, V.V.; Michor, H.; Schafler, E.; Grytsiv, A.; Bauer, E.; Rogl, P. High-ZT half-Heusler thermoelectrics, $Ti_{0.5}Zr_{0.5}NiSn$ and $Ti_{0.5}Zr_{0.5}NiSn_{0.98}Sb_{0.02}$: Physical properties at low temperatures. *Acta Mater.* **2019**, *166*, 466–483. [CrossRef]
13. Takeuchi, A.; Inoue, A. Calculations of Mixing Enthalpy and Mismatch Entropy for Ternary Amorphous Alloys. *Mater. Trans. JIM* **2000**, *41*, 1372–1378. [CrossRef]

metals

MDPI

Review

A Comprehensive Review on Hydrogen Absorption Behaviour of Metal Alloys Prepared through Mechanical Alloying

Thabang Ronny Somo [1,2], Thabiso Carol Maponya [1], Moegamat Wafeeq Davids [2], Mpitloane Joseph Hato [1,*], Mykhaylo Volodymyrovich Lototskyy [2] and Kwena Desmond Modibane [1,*]

[1] Nanotechnology Research Lab, Department of Chemistry, School of Physical and Mineral Sciences, University of Limpopo (Turfloop), Polokwane 0727, South Africa; somoronny@gmail.com (T.R.S.); thabiso.maponya@ul.ac.za (T.C.M.)

[2] Hydrogen South Africa (HySA), South African Institute of Advanced Chemistry, University of the Western Cape, Cape Town 7535, South Africa; 2235735@myuwc.ac.za (M.W.D.); mlototskyy@uwc.ac.za (M.V.L.)

* Correspondence: mpitloane.hato@ul.ac.za (M.J.H.); kwena.modibane@ul.ac.za (K.D.M.); Tel.: +27-(0)15-268-4116 (M.J.H.); +27-(0)15-268-3783 (K.D.M.)

Received: 21 March 2020; Accepted: 18 April 2020; Published: 26 April 2020

Abstract: Hydride-forming alloys are currently considered reliable and suitable hydrogen storage materials because of their relatively high volumetric densities, and reversible H_2 absorption/desorption kinetics, with high storage capacity. Nonetheless, their practical use is obstructed by several factors, including deterioration and slow hydrogen absorption/desorption kinetics resulting from the surface chemical action of gas impurities. Lately, common strategies, such as spark plasma sintering, mechanical alloying, melt spinning, surface modification and alloying with other elements have been exploited, in order to overcome kinetic barriers. Through these techniques, improvements in hydriding kinetics has been achieved, however, it is still far from that required in practical application. In this review, we provide a critical overview on the effect of mechanical alloying of various metal hydrides (MHs), ranging from binary hydrides (CaH_2, MgH_2, etc) to ternary hydrides (examples being Ti-Mn-N and Ca-La-Mg-based systems), that are used in solid-state hydrogen storage, while we also deliver comparative study on how the aforementioned alloy preparation techniques affect H_2 absorption/desorption kinetics of different MHs. Comparisons have been made on the resultant material phases attained by mechanical alloying with those of melt spinning and spark plasma sintering techniques. The reaction mechanism, surface modification techniques and hydrogen storage properties of these various MHs were discussed in detail. We also discussed the remaining challenges and proposed some suggestions to the emerging research of MHs. Based on the findings obtained in this review, the combination of two or more compatible techniques, e.g., synthesis of metal alloy materials through mechanical alloying followed by surface modification (metal deposition, metal-metal co-deposition or fluorination), may provide better hydriding kinetics.

Keywords: metal hydrides; hydrogen storage; hydriding kinetics; surface modification; mechanical alloying

1. Introduction

The physical storage of hydrogen through compressed H_2 tanks, liquid H_2 tanks and cryo-compressed tanks has been condemned, due to several limitations, including the high pressure required for compression of the hydrogen gas and a large amount of energy for liquefaction of hydrogen [1]. A convenient way to store hydrogen is through the absorption of hydrogen onto solid nanostructured materials such as metal alloys, chemical sorbents (metal organic frameworks (MOFs)) and carbon-based materials) or chemical hydrides [2]. Out of the above-mentioned chemical storage materials, metal hydrides offer a safe, volume-efficient technology aimed at hydrogen storage for reversible on-board applications due to their large hydrogen uptake capacity (7.6 wt. % by Mg-type alloys) [3,4]. There are excellent reviews, regarding MHs, prepared by equilibrium procedure for hydrogen storage application that have been published [4–6]. However, there are not a lot of reviews on the MHs prepared by non-equilibrium procedure.

It is well-documented that metallic hydrides are chemically formed from the reaction of hydrogen gas with metal or alloy according to Equation (1) [1]:

$$\text{Metal (M)} + H_2\text{ (g)} \rightarrow \text{metal hydride (MH)}. \quad (1)$$

The formation of metal hydride undergoes six distinct states of hydrogen absorption in metals: Hydrogen to metal; physisorption, which involves physical attachment of hydrogen to a metal through van der Waals interaction; chemisorption, the chemical attachment of hydrogen to metal by forming a chemical bond; subsurface hydrogen; solid solution (α-phase) and hydride (β-phase) [7]. The formation of α- and β-phases may be described by Equations (2) and (3), respectively [8]. An α-phase is formed when small amount of hydrogen occupies the interstitial sites of the host metal M, producing an H-poor solid solution, while β-phase involves saturation of the solid solution to generate a second phase:

$$M + \frac{1}{2}yH_2 \leftrightarrow MH_y \quad (2)$$

$$MH_y + \frac{1}{2}(x-y)\,H_2 \leftrightarrow MH_x. \quad (3)$$

Where M represents metal and H represents hydrogen atoms. During the formation of β-phase, the equilibrium pressure is fixed at any given temperature. According to Gibbs' phase rule ($F = C - P + 2$, in which F is degrees of freedom, C represent the components of a reaction and P is the number of phases), the number of degrees of freedom, F, is one during the interaction of metal (M) and hydrogen (H) atoms since three phases P (two solid and one gaseous) and two components C (H and M) are available [8]. The constant pressure level is referred to as the plateau pressure. During the formation, the metal atoms stretch apart to accommodate the hydrogen atoms and the physical structure of the metal atoms may also change to form a metal hydride. For example, Table 1 shows different types of metal alloys ranging from AB to AB_5, demonstrating that hydrogenation of metal is accompanied by increase in lattice parameters of the parent alloy [9–13]. The A_2B alloys were reported to possess higher hydrogen storage capacity of 3.75 and 3.04 wt. % for P6222, and Fddd phase, respectively [11]. On the other hand, AB_5 showed to have low absorption capacity of 1.43 wt. % [10]. Moreover, peaks from X-ray diffractogram may broaden and shift towards smaller 2θ values as a result of larger lattice parameters [9].

Conversely, metal hydrides release H_2 gas when heat is applied on the hydride. This process is called desorption. Hydrogen atoms (H) move from the metal to the surface of the material, combine into hydrogen molecules (H_2) and flow away as hydrogen gas. The metal atoms contract to form its original metal crystal structure [14]. The metal alloys in Table 1 can easily be synthesized by a well-known equilibrium procedure called arc or induction melting [15,16]. However, alloys prepared through this method suffer from poor hydrogenation performances and poisoning intolerance [17]. Recent findings reported that a convenient way to overcome some of these challenges is to prepare metal alloys using mechanical alloying (MA) through high-energy ball milling technique [15,18]. In this review, we provide the recent developments on hydrogenation performances of metal alloys prepared by mechanical alloying technique, comparison of MA with other non-equilibrium techniques. Moreover, we also consider the effect of combining MA with some surface modification technologies.

We anticipate that this work can provide an inspiring perspective for further research on MHs with good hydrogen storage properties.

Table 1. Lattice parameter, a, of different types of alloys before and after hydrogenation and their hydrogen storage capacities.

Type of Alloy	Structure	Alloy Lattice Parameter, a, Å	Hydride Lattice Parameter, a, Å	Storage Capacity, wt. %	References
AB	BCC	2.976	7.029	1.75	[12]
AB_2	C14	4.866	4.902	1.70	[13]
	C15	6.939	7.158	2.01	[11]
AB_5	$CaCu_5$	5.003	5.395	1.43	[10]
A_2B	(P6222)	5.205	5.463	3.75	[11]
	(Fddd)	5.284	5.411	3.04	[11]

2. Non-Equilibrium Preparation Techniques

Mechanical, kinetic and thermodynamic performances related to the formation of metal hydride are affected by structure and phases of metal or metal alloy of interest. It is currently well-known that the structure and phase of alloy materials can be controlled by synthesizing them under non-equilibrium conditions [19]. There are several processes/techniques, which are in commercial use for the synthesis of alloy materials under such conditions. Amongst such techniques, much attention has been given to mechanical alloying, melt spinning and spark plasma sintering techniques [20]. According to Scudino et al. [21], the general underlying idea in all these techniques is to synthesize alloy materials in a non-equilibrium state by "energizing and quenching". The energizing process involves the use of external dynamical force, such as melting; evaporation; irradiation; application of pressure; or storing of mechanical energy by plastic deformation on solid-state materials to put them onto a non-equilibrium state [22]. During this process, the solid-state materials change in phase to gas or liquid. In "quenching", a subsequent step after "energization", the material of interest is extinguished into a frozen state, and it is at this point where the desired structure or phase can be easily controlled [23,24]. The alloys synthesized in this sequence possess particular structural characteristics that are difficult or rare to attain by equilibrium preparation methods: Nano-crystalline grain size with a high density of grain boundaries and even lacking of long-range order (amorphous state). These structures alloy quick hydrogen uptake and excellent cycle-life behaviour [25]. Among all preparation methods that follow "energizing and quenching" mechanism, MA has been given much attention due to its user-friendly, comparatively less costly machinery and applicability to various metal alloys [19]. In this regard, mechanical alloying experimental technique is discussed with the point of interest being its effect on hydrogenation performances of metal alloy materials.

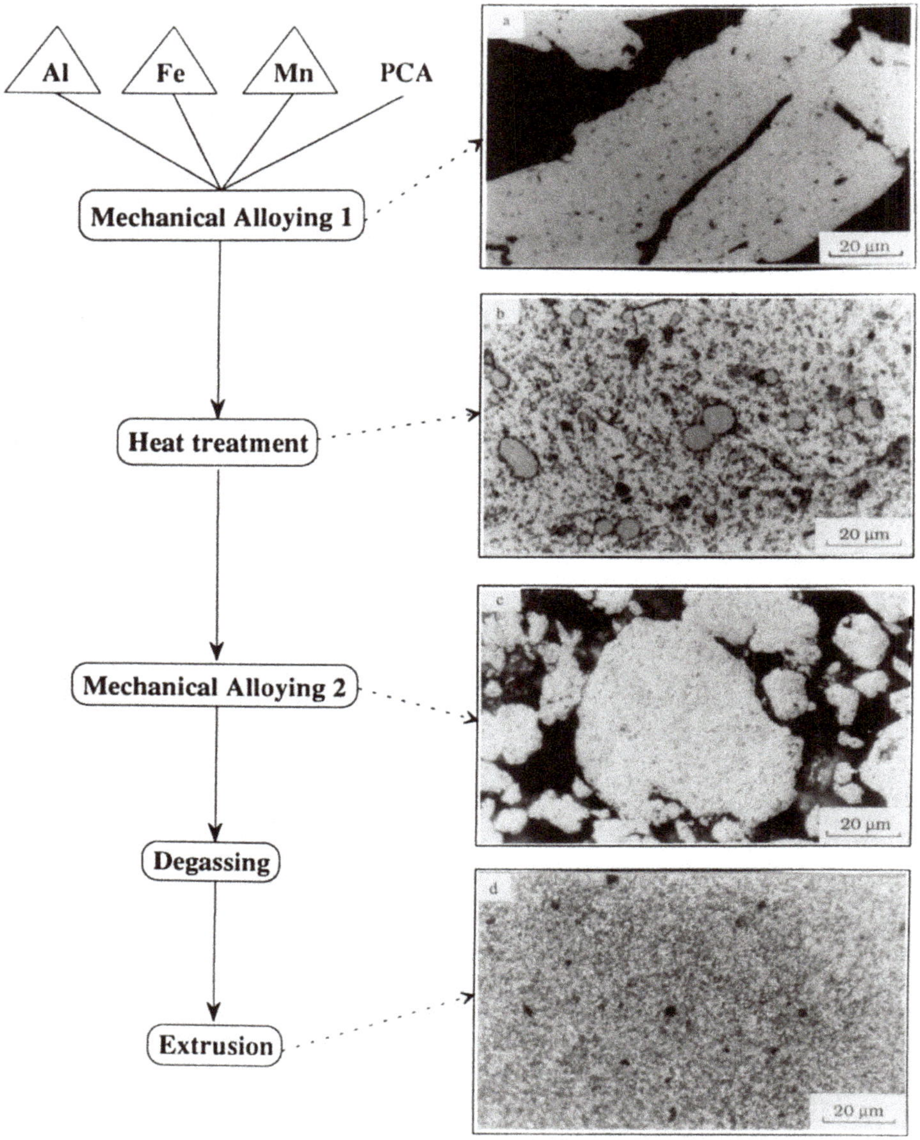

Figure 1. Process flowsheet and the microstructures developed during double mechanical alloying of an Al–5wt. %Fe–4wt. %Mn powder mixture, reproduced from [19], with permission from Elsevier, 2001.

3. Mechanical Alloying

3.1. Background

Mechanical alloying, also known as ball milling, is a solid-state powder processing technique involving repeated welding, fracturing, and rewelding of powder particles in a high-energy ball mill [19]. Repetitions of welding, as well as fracturing of particles, generate a lamellar structure within the powders. MA method is usually applied when synthesizing highly immiscible phases in a non-equilibrium state [26]. The immiscible phases that are prepared, include super-saturated solid solutions, non-equilibrium alloys that are usually amorphous in nature, metastable crystalline and quasicrystalline phases. The procedure of MA begins with blending of the powders in the correct extent and stacking the powder blend into the mill along with the grinding medium. This blend is then milled for the ideal time allotment until a consistent state is achieved when the arrangement of each powder particle is equivalent to the extent of the elements in the initial powder blend. The milled powder is then united into a mass shape and exposed to heat, in order to get the ideal phase and structure. Thus, the vital components of MA procedure are the raw materials, mill and procedure parameters. This procedure was described by Shang et al. [27]. Ares et al. [25] observed that diffraction peaks from the X-ray diffractogram (XRD) of Mm (Ni, Mn, Al, $Co)_{5.2}$ became broader with milling time. An example of MA process is depicted in Figure 1 (synthesis of Al–5wt. %Fe–4wt. %Mn blend), with scanning electron microscopic (SEM) images provided for all the steps. The double mechanical milling, as observed in the figure is necessary to refine the intermetallics and decrease the particle size of the blend [19]. MA has also been utilized to synthesize AB (FeTi), A_2B (Mg_2Ni) and AB_5 (binary $LaNi_5$) alloys [28]. In most cases, the rate of hydrogenation has been greatly enhanced when compared with that of materials prepared through conventional equilibrium methods, but due to the amorphous nature of milled alloys, a loss of hydrogen content was observed [25]. MA provides various advantages, including smaller restrictions with respect to composition and works much better when applied to A_2B alloys [29]. Moreover, through the use of MA method, nano-crystalline alloys are obtained, starting either with a pre-prepared alloy or with a mixture of pure elemental metals [28]. Nano-crystalline alloys that are prepared through MA technique possess better activation compared to their polycrystalline counterparts [28]. For example, Polycrystalline FeTi is activated by heating up to temperatures above 673 K for one to two hours followed by annealing at 7 bar pressure and its hydrogenation requires a high pressure of 40–70 bar [30]. However, with a mechanically-alloyed nano-crystalline FeTi alloy, a single vacuum heating at 673 K for 30 min is sufficient to obtain full hydrogen absorption cycles [30].

3.2. Mg-Based Alloys

3.2.1. Structural Characteristics of Mechanically Milled Mg-Based Alloys

The crystal structure of a metallic Mg is hexagonal with lattice parameters a = 3.2094 Å and c = 5.2108 Å and a space group of P63/mmc [31]. Formation of MgH_2 hydride consists of two steps, i.e., preparation of Mg nanoparticles, followed by hydrogen absorption to form the metal hydride. A traditional method for preparing Mg nanoparticles is gas-phase condensation method, which is much better discussed in [32]. As depicted by Figure 2a, hydrogenation of Mg nanoparticles tempers with XRD diffraction pattern of the parent nanoparticles [33]. Diffraction peaks at $2\theta \approx 48°$, 34° and 32° disappear, while several new peaks develop, forming tetragonal magnesium hydride (β-MgH_2) phase [33]. To illustrate the difference between traditionally synthesized MgH_2 hydride and those synthesized through mechanical/ball milling, XRD diffraction patterns of MgH_2 and as-milled hydrided Mg-Al alloy are compared in Figure 2a,b, respectively. For as-milled Mg-Al alloy, Li et al. [31] reported four stable phases: the fcc solid solution of magnesium in aluminum, the hcp solid solution of aluminum in magnesium, the β-phase (Mg_2Al_3) and the γ-phase ($Mg_{17}Al_{12}$). Such rewarding characteristic gives rise to excellent hydriding behavior of as-milled Mg-based alloys.

In general, the morphology and particle size distribution are also important structural parameters, carrying a huge responsibility for surface physisorption of molecular hydrogen (H_2), surface dissociative chemisorption, surface absorption, hydrogen diffusion and phase transformation [34]. These two parameters are well-interpreted using scanning electron microscopic (SEM) images. For Mg-based alloys, mechanically milled alloys show faster hydrogen absorption kinetics and larger uptake capacity, compared to their unmilled counterparts and one of the effective factors contributing to this, is the reduction of particle sizes of the parent materials during mechanical alloying [35]. According to the literature, small particle sizes allow easy mobility of molecular hydrogen and quick dissociation into atomic hydrogen on the surface of a material [33]. As shown in Figure 3, all the five SEM images of milled Mg-based alloys constitute of smaller particles than that of traditionally sysnthesized MgH_2, [33]. Therefore, their hydrogen uptake performance is expected to be superior to that of MgH_2. This is further discussed in the next section.

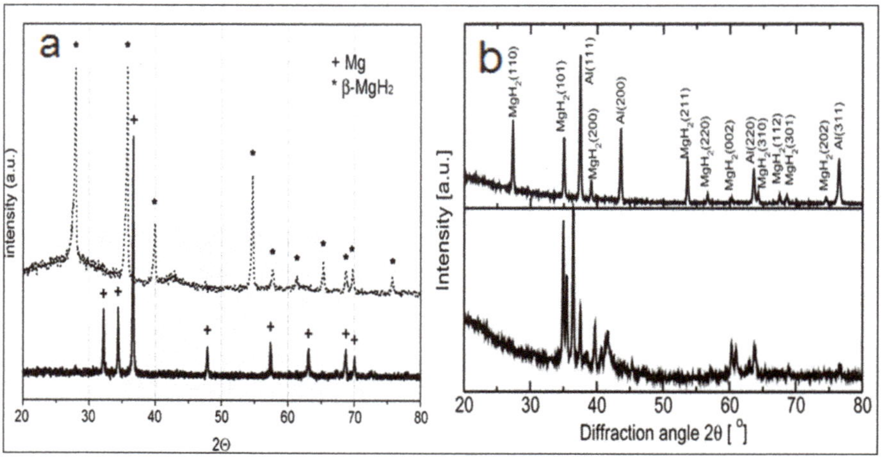

Figure 2. XRD diffraction patterns of (**a**) nanocrystalline Mg and MgH_2, and (**b**) Mg-Al alloy in the hydrogenated state (top) and dehydrogenated state (bottom). Figures reproduced from [33], and [31], respectively, [31] with permission from Elsevier, 2008; [33] with permission from Elsevier, 2007.

Figure 3. The SEM images of the (**a**) MgH_2, (**b**) as-milled MgH_2, (**c**) as-milled MgH_2 + 10wt. % $CeCl_3$, (**d**) as-milled Mg–10 $FeTiO_3$, (**e**) as-milled Mg:Al (90:10) and (**f**) as-milled MgH–5 at. % V. Figures (**a–c**) were reprinted with permission from [36], with permission from Elsevier, 2016 (**d**) reproduced from [37], with permission from Elsevier, 2015; (**e**) reproduced from [38], with permission from Elsevier, 2000 and (**f**) reproduced from [39], with permission from Elsevier, 1999.

3.2.2. Hydrogen Absorption Behaviour of Mechanically Alloyed Mg-Based Alloys

Mg-based alloys are reported to possess energy density of 9 MJ/kg and the highest maximum hydrogen storage capacity (7.6 and 3.6 wt. % for MgH_2, and Mg_2NiH_4, respectively) compared to all hydride-forming materials [40,41]. The hydride stability of this alloy is very high, and therefore, a high dissociation temperature of around 561 K is required to transform MgH_2 hydride into Mg and H_2 gas [42]. Furthermore, the hydriding kinetics of MgH_2 hydride is relatively slow because the dissociation of molecular hydrogen into hydrogen atoms is difficult on the surface of Mg metal [43]. Several reports are found in the literature concerning synthesis and hydrogen properties of Mg-based alloys synthesized by mechanical alloying [3,44–46]. Another interesting study is Andreasen [31]. The authors'work was based on the comparison of hydrogen kinetics, presented in Figure 4a, of un-milled Mg and those of ball milled Mg-Al compounds with varying Al content [31]. The kinetics increased gradually with an increase in Al content due to easy dissociation of hydrogen molecules on the surface of Al, as a result of negligible activation barrier for hydrogen dissociation for this metal. It is, therefore, understood that hydrogen molecules split into atomic hydrogen on the surface of Al before diffusing and attaching on the active sites of Mg, which has higher hydrogen sticking ability compared to Al [31]. Lototskyy et al. [37] reported the effect of synthesizing different Mg-based composites varying the amounts of multi-walled carbon nanotubes, activated carbon, $FeTiO_3$ and graphite using mechanical alloying at a rotation speed of 500 rpm for a period of 120 min on hydrogenation behaviour of Mg, Figure 4b. The authors discovered that hydrogenation behaviour (both the uptake capacity and kinetics) of Mg mechanically alloyed with carbon-based materials are exceptional. Ref. [46] reveals that the better hydrogenation behaviour possessed by milled alloys can be attributed to their combination of tetragonal (a = 4.52 Å, c = 3.02 Å) and orthorhombic (a = 4.48 Å, b = 5.40 Å, c = 4.90 Å) phases as major phases, since these two phases are associated with small particle size and a presence of strain in the particles. Similar trends of hydrogen capacity and kinetics were observed by Lototskyy et al. [47] and Williams et al. [48] when they assessed Mg-10(FeV)-5MWCNT and Mg-5Pd composites through mechanical alloying, respectively. Tables 2 and 3 represent a compiled data of literature based on hydrogenation behaviour of Mg-based alloy materials synthesized through ball milling and mechanical alloying.

Table 2. Thermodynamic and kinetic properties of ball-milled MgH_2 doped with catalytic additives.

Catalyst Family	Catalyst Additive	ΔH_{des}, kJ/mol H_2	Hydriding Kinetics, min^{-1}	Hydrogen Capacity, wt. %	References
No catalyst	Undoped MgH_2	76	1.4×10^{-3}	7.66	[52,57]
Metal	MgH_2+10 wt. % Ni	75	–	6.89	[49]
	MgH_2+10 wt. % Co	71	–	5.62	[49]
Metal halides	MgH_2+10 wt. % $CeCl_3$	75.7	0.05	6.32	[36]
	MgH_2+NbF_5	79.7	3.42	6.40	[53]
	MgH_2+ZrF_4	76.4	3.2	6.44	[53]
Hydrides	MgH_2+$NaAlH_4$	–	0.28	3.10	[49]
Metal oxide	MgH_2+$SrTiO_3$	–	0.02	6.63	[50]
	MgH_2+1 mol. % Nb_2O_5	74	0.22	5.45	[51]
Nano-sized alloys	MgH_2+10 wt. % Zr_2Ni	–	0.52	5.12	[55]
	MgH_2+FeNb	–	0.35	5.55	[56]
Carbon materials	MgH_2+5 wt. %SWNT	–	1.34	6.72	[54]

ΔH_{des} = Enthalpy of desorption*SWNT = Single-walled carbon nanotube.

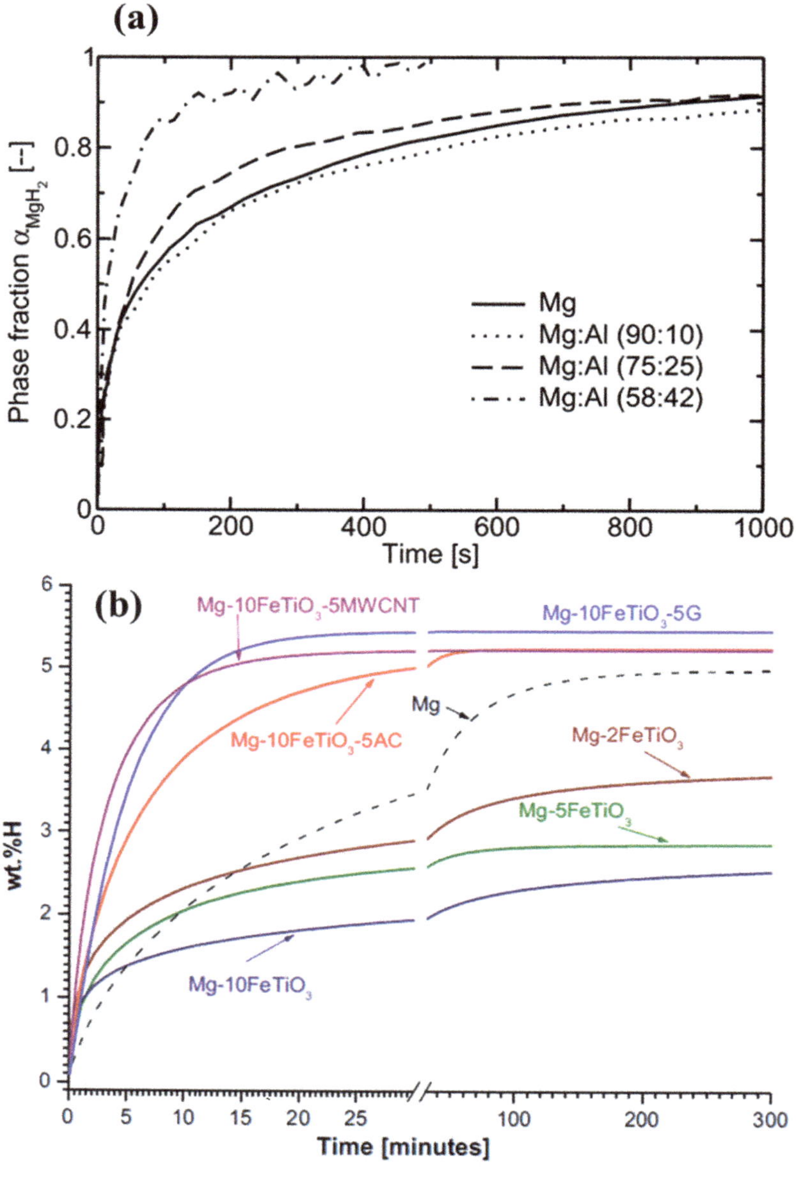

Figure 4. (**a**) Hydrogenation kinetics of pure un-milled Mg compared with Mg-Al compounds with varying Al content ball milled for 20 h. Hydrogenation performed at T = 400 °C and a hydrogen pressure of 38 bar. Data extracted from ref. [31], with permission from Elsevier, 2008. (**b**) Hydrogenation kinetics of Mg and its carbon-based composites (15 bar and H_2/250 °C), reproduced from [37], with permission from Elsevier, 2015.

3.2.3. Ball-Milled MgH_2 Doped with Catalytic Additives

In the previous section, much attention is given to mechanically milled Mg-based alloys where a pure milled/un-milled Mg metal is admixed with other several materials, such as $FeTiO_3$ before their interaction with hydrogen is subsequently explored. Here, we discuss the influence of catalytic effect of doping ball milled MgH_2 with different catalystic additives. Previous experiments indicate clearly that admixing, deposition or doping of catalysts and additives could greatly enhance thermodynamic property and de/rehydrogenation kinetic parameters of MgH_2. In these studies, metal [49], metal oxide [50–52], metal halides [36,53], metal hydride [49], carbon materials [54] and nanosized alloys [55,56] were used as the catalysts and additives. Table 2 summarizes thermodynamic and kinetic effects of these catalysts. The materials in the table were prepared through ball milling technique.

We can conclude from the table that doped MgH_2 materials, prepared through mechanical milling, enjoy superior hydrogenation kinetics, compared to undoped materials. However, this particular excellent improvement significantly reduces the maximum absorption capacity of the system. For metal halide family, particularly MgH_2+10wt. %$CeCl_3$, the loss of capacity can be related to in situ formation of CeMg alloy and $MgCl_2$, which seem to be good catalysts for faster absorption but poor gas absorbers [53]. On the other hand, drastic decrease of absorption capacity revealed by MgH_2+SWNT composite is suggested to come from structure destruction of the SWNT during milling [54]. Table 2 also shows that composites which resulted in major decrease in hydrogen capacity are those with a dopant of 10 wt. % or more. We therefore suggest preparation of MgH_2 composites with dopants that are less than a concentration of 10 wt. %. Moreover, milling period should also be taken into consideration when these materials are formed, more especially when carbon materials are in abundance as they seem to be sensitive to milling conditions. The difference between enthalpy of desorption (ΔH_{des}) of the materials is negligible and as thus thermodynamic property of MgH_2 remain unchanged upon doping.

3.3. *AB_5-Type Alloys*

3.3.1. Structural Characteristics of Mechanically Milled AB_5-Type Alloys

Another interesting class of metal alloys is AB_5. The A site in AB_5 represents one or more strongly hydride-forming elements while B site represents one or more weakly hydride-forming elements, and usually helps to dissociate the H_2 molecules on the surface of alloy material [58]. The A site is usually occupied by La, Ca or rare earth element while Ni, Cu, Co, Pt or Fe fill the B site [59]. AB_5 alloys are known to have a $CuCa_5$-type hexagonal crystal structure which normally belongs to space group P6/mmm (#191) [60,61]. Singh et al. [62] evaluated the effect of milling $MmNi_{4:6}Fe_{0:4}$ alloy, where Mm refers to mischmetal, varying the milling period from 10, 20 and 30 min, at a milling speed of 200 rpm. As shown by Figure 5, mechanical milling of this alloy results in slight shift of diffraction peaks towards lower 2θ values, reflecting increase in lattice parameters, and peak width broadening. Peak width broadening is a reflection of decrease in particle size. In general, the lattice structure of $MmNi_5$-type alloy consists of two stacked planes at $\zeta = 0$ and $\zeta = \frac{1}{2}$, and the first plane contains 1a and 2c, while the latter plane contains 3g crystallographic sites. Both Mm and Ni atoms occupy the 1a and 2c sites, while the 3g sites are occupied by Ni atoms only. Therefore in the presence of Fe to form $MmNi_{4:6}Fe_{0:4}$ alloy, Fe atoms replace Ni on the 3g sites [63]. Electronic configuration of Fe and that of Ni are very close ($3d^6 4s^2$ for Fe and $3d^8 4s^2$ for Ni) and as a result covalent character, which is stronger than ionic character, is expected in the Ni/Fe-bearing planes. Thus, during mechanical milling of an alloy material with such character, strained platelet shaped particles are expected to form [64]. Such particles allow smooth movement of hydrogen during sorption better than fine powdered particles. This statement is supported by Table 3 as it is observed from the table that materials that were milled for longer period of time are associated with lower hydrogen uptake capacity compared to their counterparts that were milled for shorter period. This is due to the mechanical energy that is being pumped into the alloy matrix during milling, which breaks the particles, while forming strain in the matrix, and thus, for longer milling periods fractured and much finer powders, which are not recommended may result [63,64].

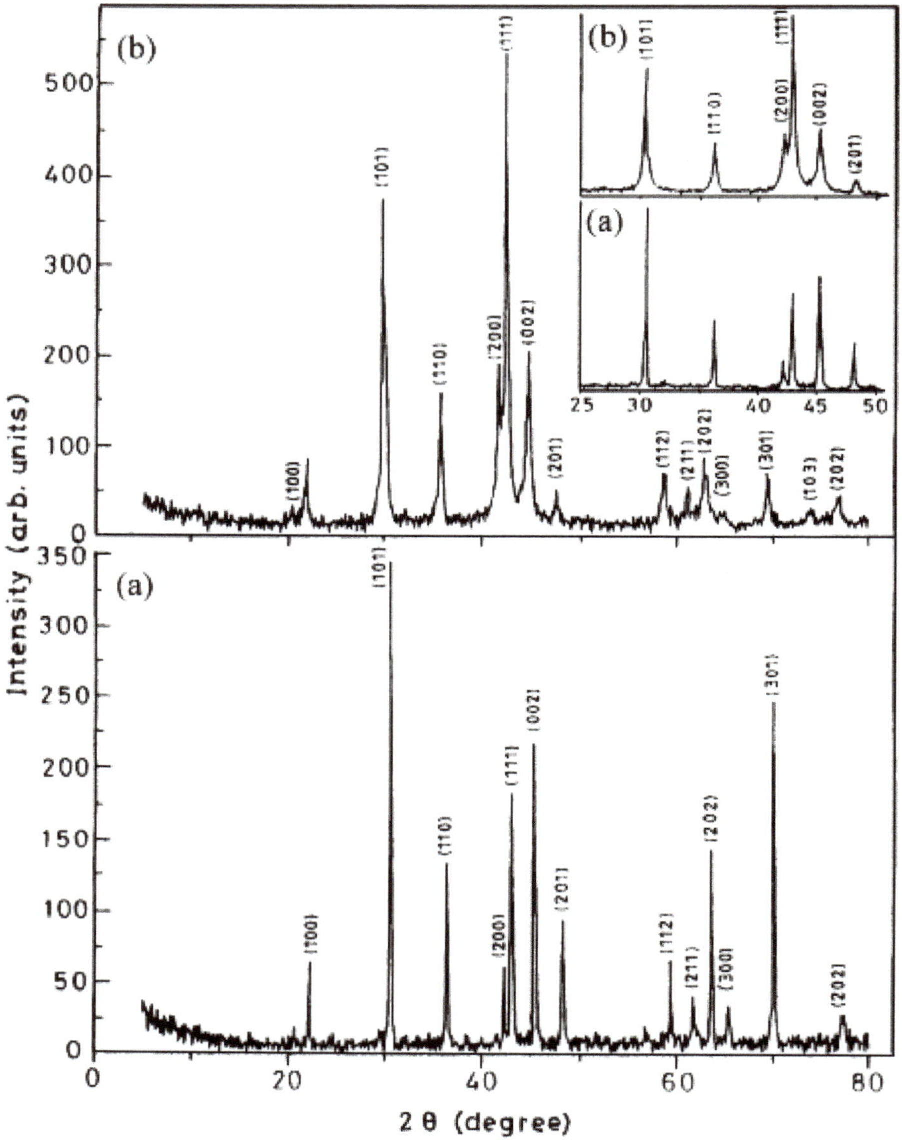

Figure 5. X-ray difractograms of the unmilled; (**a**) and mechanically pulverized; (**b**) $MmNi_{4:6}Fe_{0:4}$ alloys and the inset picture focus on broadening of full width at half maximum (FWHM). reproduced from [61], with permission from Elsevier, 2004.

Table 3. Hydrogenation properties of metal alloys synthesized by mechanical alloying.

Alloy type	Alloy material	Milling conditions (rotation speed and milling time)	Particle size distribution, μm	Hydriding kinetics, min^{-1}	Hydrogen capacity, wt. %	References
			Mg-based alloys			
As-cast alloy	Mg	–	48–70	1.4×10^{-3}	7.66	[57]
As-milled alloys	Mg	500 rpm for 120 min	<0.05	0.04	4.98	[37]
	$Mg\text{-}10FeTiO_3$	500 rpm for 120 min	<0.05	0.07	2.71	
	$Mg\text{-}10FeTiO_3\text{-}5MWCNT$	500 rpm for 120 min	<0.05	0.31	5.22	
	Mg-10(FeV)	500 rpm, for 300 min	0.001–0.01	0.02	6.96	[47]
	Mg-10(FeV)-5MWCNT	500 rpm, for 300 min	~0.01	0.02	6.69	
	Mg-0.5Pd	500 rpm, for 360 min	0.05–0.1	0.04	5.13	[48]
	Mg-5Pd	500 rpm, for 360 min	<0.01	0.02	5.56	
			AB_2-type alloys			
Alloy type	**Alloy material**	**Milling conditions (rotation speed and milling time)**	**Particle size distribution, μm**	**Hydriding kinetics, min^{-1}**	**Hydrogen capacity, wt. %**	**References**
As-cast alloy	$Ti_{0.5}Zr_{0.5}MnCr$	–	15–29	0.04	1.79	[65]
As-milled alloys	$Ti_{0.5}Zr_{0.5}MnCr$	Milled for 5 min	~8	0.13	1.82	
		Milled for 30 min	3–6	0.05	1.65	
		Milled for 60 min	0.5–1.7	0.06	1.15	
As-milled alloys	$Mg(Ni_{0.75}Mn_{0.25})_2$	350 rpm for 600 min	<10	0.25	0.43	[66]
	$Mg(Ni_{0.1}Mn_{0.9})_2$	350 rpm for 600 min		0.14	0.93	
	$MgMn_2$	350 rpm for 600 min		0.1	1.21	
			AB_5-type alloys			
Alloy type	**Alloy material**	**Milling conditions (rotation speed and milling time)**	**Particle size distribution, μm**	**Hydriding kinetics, min^{-1}**	**Hydrogen capacity, wt. %**	**References**
As-cast alloy	$MmNi_5$	–	5–50	0.24	1.20	[67]
As-milled alloy	$MmNi_5$	200 rpm for 180 min	1–2	0.28	1.42	
As-cast alloy	$MnNi_{4.6}Fe_{0.4}$	–	~9.58	–	1.50	[62]
As-milled alloys	$MnNi_{4.6}Fe_{0.4}$	200 rpm for 10 min	–	–	2.00	
		200 rpm for 30 min	~1.62	–	1.82	
		300 rpm for 10 min	–	–	1.91	
		300 rpm for 30 min	–	–	1.21	

To be more accurate, not only particle size gives precise measurement of hydrogen uptake capacity by a material, but it also provides the nature of agglomeration, as well as homogeneity of the particles [62]. Short milling process yields non-homogeneous microstructural (wide distribution of nano-crystallite sizes and amorphous phase) and morphological alloy powders [25]. On the other hand, Parker et al. [35] recommends defects, fractures and porosity as faster absorption kinetics agents. However, it is very challenging to fully release hydrogen from a porous and fractured material. Figure 6 depicts SEM images of un-milled $MmNi_5$ alloy and those of $MmNi_5$ milled with different transition metals at concentration of 2.0 wt. %. for un-milled alloy, its structure lacks strain and consists of very large particles, and thus, hydrogen diffusion is difficult for such material [67].

In contrast to $MmNi_{4:6}Fe_{0:4}$ alloy, neither covalent nor ionic bonding is expected between the bulk alloy $MmNi_5$ and the added transition metals (Co, Ni, Mn and Fe), since there was no substitution or melting during mechanical milling, thus, a physical attachment was involved [67]. As a result, the extent to which the particles of each alloy crush and become reduced in size during milling depends largely on ductility and hardness of the transition metal used. According to Mohs scale of metal hardness, Mn has Mohs hardness of 6, followed by Co with hardness of 5 and the last two are Fe and Ni both with hardness of 4.5 and 4, respectively [68]. Therefore, during ball milling, it is difficult to fracture and form finer powders with Mn particles than with Ni particles.

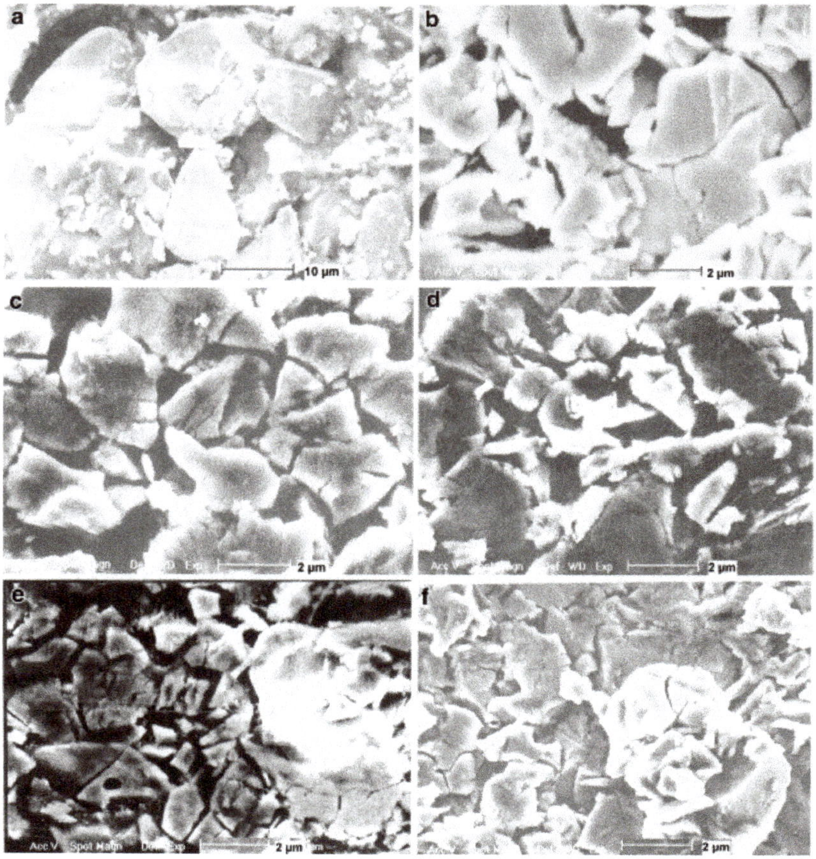

Figure 6. SEM micrographs of (**a**) bulk $MmNi_5$ alloy; (**b**) ball-milled $MmNi_5$ alloy; (**c**) $MmNi_5$ alloy ball-milled with Co; (**d**) Ni; (**e**) Mn and (**f**) Fe transition metals at concentration of 2.0 wt. %, obtained from [60].

3.3.2. Hydrogen Absorption Behaviour of Mechanically Alloyed AB_5-Type Alloys

These metal hydride alloys are typified by $LaNi_5$ and they can be easily activated for hydrogen absorption [69]. A reaction between these alloys and hydrogen at a room temperature and pressure above equilibrium plateau pressure results in a ternary metal hydride [69]. Of 37 voids available within $LaNi_5$ lattice structure, only 6 voids are occupied by hydrogen atoms to form $LaNi_5H_6$ ternary hydride [70]. It is notable that the hysteresis effect in AB_5-type alloy is relatively small as compared to other low temperature systems with a large and distinct miscibility gap [59]. Mechanical alloying has also been employed to synthesize AB_5 alloys [25,62,67,71]. For example, $MmNi_5$ material mechanically alloyed at 200 rpm for milling time of 180 min exhibits faster hydrogenation kinetics as compared to arc-melted $MmNi_5$ material [67]. Furthermore, Figure 7a,b present the absorption and desorption kinetic curves of $MmNi_5$ alloy ball-milled with transition metals at concentration of 2.0 wt. %, respectively, showing reversible hydrogen absorption and desorption between 1.0 to 1.6 wt. % [67]. Singh and co-workers [62] observed that longer milling of $MnNi_{4.6}Fe_{0.4}$ alloy material improved hydriding kinetics, but this elevated kinetics behaviour comes at a cost of maximum uptake capacity as it deteriorated with increasing milling time. On the same note, an increase in rotation speed from 200 rpm to 300 rpm at constant milling time resulted in a reduction of hydrogen capacity. Similar trend was observed by Ares and co-workers [25] and they summarized that the trend is a result of extra hydrogen-trapping sites in amorphous domains, exhibited by short-milled alloys. For $LaNi_5$ alloy, it is reported that ball-milling convert a part of $LaNi_5$ to a non-absorbing state and as a result poor hydrogen uptake capacity is observed [71]. It was seen that there is quite a number of published work found in the literature for mechanical alloying of AB_5-type alloys [62,67]. However, understanding the general extent to which mechanical alloying enhance the ability of these type of alloys to absorb hydrogen, as compared to their traditionally or arc-melted counterparts is still a gap. Firstly, to our knowledge no proper measurements of the hydrogen sticking or diffusion probabilities on either un-milled or milled AB_5-type particles have been documented. The second issue is that mechanical alloying tempers with several physical properties that are vital for hydrogen sorption performance; these include expansion of the unit cell volume, strain and homogeneity of particles, particle size distribution, distance between adjacent voids and several more.

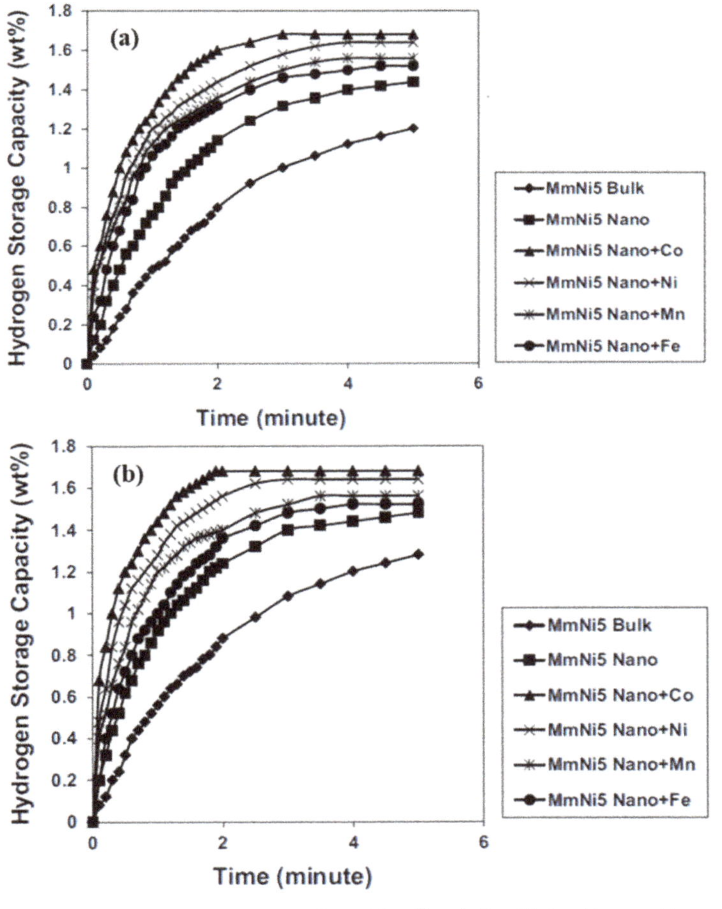

Figure 7. (**a**) Absorption kinetic curves of $MmNi_5$ alloy ball-milled with transition metals at concentration of 2.0 wt. %, and (**b**) Desorption kinetic curves of $MmNi_5$ alloy ball-milled with transition metals at concentration of 2.0 wt. %. Reprinted from [67].

3.4. AB_2-Type Alloys

3.4.1. Structural Characteristics of Mechanically Milled AB_2-Type Alloys

In the AB_2 Laves phase alloy, metal A is usually Ti or Zr, and metal B is another transition metal including V, Cr or Mn. These alloys crystallize in Lave phase structures [72,73]. Three types of laves phases; namely: hexagonal C14 phase ($MgZn_2$), hexagonal C36 phase ($MgNi_2$) and cubic C15 phase ($MgCu_2$) represent AB_2- type alloys [72]. In comparison, hydrogenation performance of hexagonal C36 phase has been reported to be poor [73], and it is not reviewed here. C15 and C14 phases (corresponding to atomic ratio of $R_A/R_B = 1.225$) have shown better performance as hydrogen absorbents [73]. Based on their structural stability, which is dependent on the atomic ratio, electronegativity and valence of the elements used [74,75], hexagonal C14 $ZrCr_2$ Laves phases are more stable compared to cubic C15 [76].

Figure 8 represents example of powder diffraction patterns for C14 Laves phase type alloys. As seen with Mg-based and AB_5-type alloys discussed earlier in this review, longer milling periods results in reduction of crystallite size and particle size. Furthermore, several other factors such as disappearance of peaks (between $2\theta \approx 48$ and 100 in Figure 8), as well as peak broadening occur. To some extent, intense change of these factors may result in deterioration of hydrogen performance.

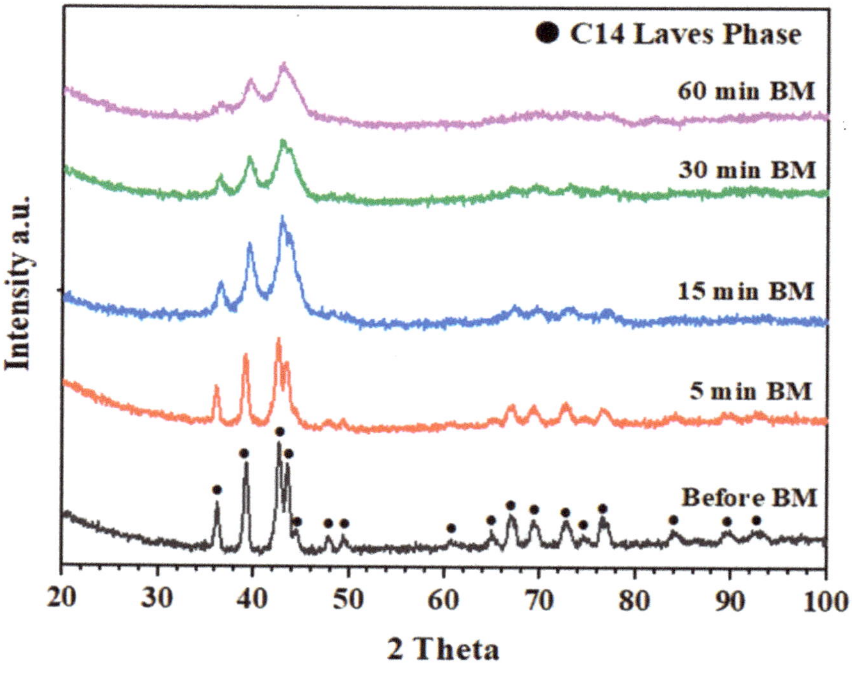

Figure 8. Powder diffraction patterns of alloy $Ti_{0.5}Zr_{0.5}MnCr$ before and after 5, 15, 30 and 60 min of ball milling, reproduced from [65], with permission from Elsevier, 2019.

3.4.2. Hydrogen Absorption Behaviour of Mechanically Alloyed AB_2-Type Alloys

The reported hydrogen storage of AB_2-type alloys is nearly 2 wt. %, with low hysteresis [75]. But binary AB_2-type alloys prepared by equilibrium techniques suffer from poor surface activity and poisoning intolerance [77]. As a result, efforts through synthesizing AB_2 alloys utilizing non-equilibrium techniques such as mechanical alloying have been exploited [66]. For example, $Ti_{0.5}Zr_{0.5}MnCr$ material prepared through mechanical alloying for a milling period of 5 min was reported to exhibit larger hydrogen absorption capacity compared to as-cast $Ti_{0.5}Zr_{0.5}MnCr$ and materials milled much longer (Table 3 and Figure 9) [65].

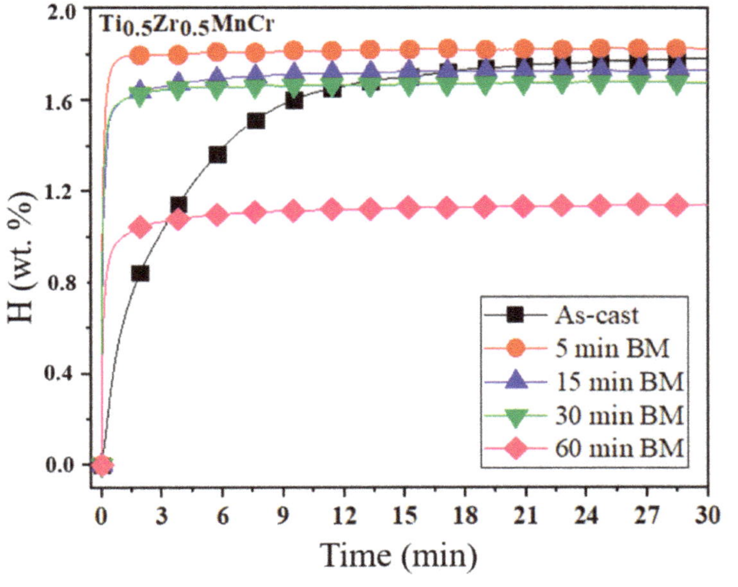

Figure 9. Hydrogenation kinetics of $Ti_{0.5}Zr_{0.5}MnCr$ alloys before and after ball milling for 5, 15, 30 and 60 min, reproduced from [58], with permission from Elsevier, 2019.

However, those that are milled longer than 5 min exhibit faster hydriding kinetics and poor hydrogen absorption capacity compared to as-cast alloy [65]. Based on the literature data contained in this document, we understand that fractured fine, smaller particle sizes and amorphous nature, which are obtained through milling time above 25 min favour faster hydrogenation kinetics. Whereas combination of large to moderate particles and nano-crystalline structure, obtained during short milling time of less than 10 min favours large hydrogen absorption capacity.

3.5. Comparison between Hydriding Kinetics of Materials Prepared by MA with Other Techniques

Apart from mechanical alloying, melt spinning and spark plasma sintering techniques are also prevailing synthesis methods and thus we opted to make some comparisons between these three methods. Melt-spinning technique is a useful method that is used to change structures of as-cast alloys to amorphous structure. This method improves the amorphous nature of materials by reducing their grain size [44]. The only problem related to this method is the poor cycle stability of the melt-spun alloys, due to the disappearing of the metastable structures generated by melt-spinning [78]. Another attracting technique is spark plasma sintering (SPS), due to its ability to easily synthesize metal powders, with different melting temperatures within just 5–10 min [79]. This technique involves generation of spark plasma through direct pulse current and working concurrently with a uniaxial pressure. Sintering takes place at low temperatures in a small period of time and this makes the alloys to have low porosity [80]. SPS offers numerous favourable advantages over customary sintering procedures such as atmospheric furnace. For instance, the heating rate around 1273 K per min can be accomplished [79]. Further focal points are the lower sintering temperature, short holding time, no need of pre-compaction and shorter sintering time results in the likelihood to sinter nanometric powder to near theoretical density [81]. Although this technique produces alloys with high purity surfaces, it is known that the hydriding kinetics of composites, prepared by SPS, is greatly improved only when hydrogen absorption is conducted at high temperatures (573 K) [79].

Huang et al. [82] reported that alloys with amorphous surfaces absorb hydrogen faster compared to their crystalline counterparts. Schwarz and Johnson [83] recommended that two conditions should be fulfilled for an amorphous phase to form from a mixed elemental powder blend. The first condition is that the metals involved must have a vast negative heat of blending in the liquid state. According to a study conducted by Termsuksawad et al. [84], metals that have large negative heat of mixing exhibited a complete amorphous phase, while those with smaller negative heat displayed partial amorphous nature. Second, the condition is that diffusion rates of such metal systems must differ greatly. Metal alloy materials with nano-crystalline and/or amorphous phases are easily obtainable through several preparation techniques, including mechanical milling, spark plasma sintering and melt spinning [85,86]. There are numerous published papers supporting the above statement, and we combined some of those findings in Table 4. All the as-milled, as-spun and sintered alloy materials in Table 4 are reported to exhibit amorphous phases, and they are compared with their as-cast crystalline counterparts. According to the table, mechanical alloying of as-cast materials usually has a negative impact on the maximum hydrogen absorption capacity as all materials showed drastic decrease in their respective hydrogen content, except for $Ti_{0.5}Zr_{0.5}MnCr$ alloy. Similar to other authors, Khajavi et al. [65] rationalized a loss in hydrogen capacity by deducing that milling process decreases the crystallite size, resulting in peak broadening. Nonetheless, milling significantly enhanced the absorption kinetics, and this is indicated by exceptional rate constants associated with all the as-milled counterparts in Table 4. Almost all the rate constants presented in the table were estimated from kinetic curves using chemisorption model [65] as shown by Equation (4),

$$\alpha = kt \tag{4}$$

where α is the reacted fraction, t is hydrogenation time and k is rate constant. Metastable alloys (as amorphous, nano-crystalline, quasicrystalline, polycrystalline) can be prepared through melt spinning technique, since it allows quick quenching of melted metals. Most as-spun materials reported in this review have amorphous (examples being $Mg_{11}Y_2Ni_2$ and $Mg_3LaNi_{0.1}$) and nano-crystalline phases (examples being Mg-10Ni-2Mn and $Mg_2Ni_{0.9}Co_{0.1}$). On a contrary to mechanical alloying, melt spinning technique enhances both the hydrogen storage capacity and the kinetics (Table 4). The speedy hydrogenation kinetics for as-spun materials can be related to fine particle size and the crystal defects shaped as a result of rapid solidification, which facilitates swift nucleation and diffusion of hydrogen [23]. The enormous quantity of interfaces and grain boundaries created within the as-spun alloy offers easy access for hydrogen diffusion and facilitate easy absorption of hydrogen [23]. Another interesting technique featured in this review is spark plasma sintering. In the same way as MA and melt spinning, SPS-synthesized alloy materials exhibit an amorphous/nanocrystallite phase [87,88]. The absorption kinetics and capacity of sintered materials are affected by the amount (percentage), the type of material, and sintering conditions (temperature and pressure) used for SPS process. For example, hydrogen content of metallic Mg was reduced extremely from 7.26 wt. % to 2.52 wt. % upon blending with 77% $V_{77.8}Zr_{7.4}Ti_{7.4}Ni_{7.4}$ through SPS while hydrogen content of $V_{35}(Ti,Cr)_{65}$ sintered with $ZrMn_2$ remained almost the same [87,88]. Interestingly, the rate constants of sintered alloy materials in Table 4 are better than those of as-cast materials. This is a clear indication that the SPS method has a positive effect on hydrogen absorption kinetics. Thus, it can be seen that all these three techniques are good candidates for improvement of hydrogen absorption kinetics, but alloys prepared using SPS technique in most cases have low hydrogen capacity. Hence, we take a further comparison between the prevailing techniques–MA and melt spinning.

Table 4. Comparison of hydrogenation kinetics and capacity of as-cast alloys and those prepared by MA, SPS and melt spinning.

Mechanical Alloying/Milling						
Material	**Rate constant (min^{-1})**		**Hydrogen absorption capacity (wt. %)**		**Conditions**	**References**
Metal alloy material	As-cast	As-milled	As-cast	As-milled		
Mg_2Ni	0.06	0.15	2.91	2.60	473 K, 1.0 MPa	[66]
TiFe+4 wt. % Zr (60 min milling)	9.0×10^{-3}	0.01	1.62	1.27	273 K, 4.5 MPa	[89]
TiFe+4 wt. % Zr (5 min milling)	9.0×10^{-3}	8.5×10^{-3}	1.62	1.41	273 K, 4.5 MPa	[89]
$Ti_{0.5}Zr_{0.5}MnCr$ (60 min milling)	0.04	0.05	1.75	1.05	273 K, 2.0 MPa	[65]
$Ti_{0.5}Zr_{0.5}MnCr$ (5 min milling)	0.04	0.05	1.75	1.82	273 K, 2.0 MPa	[65]
Melt-spinning technique						
Material	**Rate constant (min^{-1})**		**Hydrogen absorption capacity (wt. %)**		**Conditions**	**References**
Metal alloy material	As-cast	Melt-spun	As-cast	Melt-spun		
$Mg_{11}Y_2Ni_2$	0.90	0.52	3.61	3.89	523 K, 3.0 MPa	[90]
$Mg_3LaNi_{0.1}$	0.25	0.60	2.73	2.90	573 K, 4.0 MPa	[24]
Mg-10Ni-2Mn	0.01	0.02	4.67	5.09	598 K, 1.0 MPa	[23]
$Mg_2Ni_{0.9}Co_{0.1}$	9.0×10^{-3}	0.02	2.38	3.00	473 K, 1.5 MPa	[91]
Spark plasma sintering						
Material	**Rate constant (min^{-1})**		**Hydrogen absorption capacity (wt. %)**		**Conditions**	**References**
Metal alloy material	As-cast	Sintered	As-cast	Sintered		
Mg sintered with 77% $V_{77.8}Zr_{7.4}Ti_{7.4}Ni_{7.4}$	1.7×10^{-3}	6.7×10^{-3}	7.26	2.52	573 K, 3.0 MPa	[87]
$V_{35}(Ti,Cr)_{65}$ sintered with $ZrMn_2$	0.01	0.04	2.86	2.89	303 K, 0.6 MPa	[88]
Mg sintered with 30% $ZrMn_2$	1.7×10^{-3}	3.3×10^{-3}	7.26	6.34	573 K, 3.0 MPa	[87]

Zhang et al [44] and Wu et al [92] made some efforts to prepare and compare structures and hydrogenation behaviour of as-milled and as-spun alloy materials. Their findings are summarized in Table 5. From the table, it can be seen that the hydrogen absorption capacity of Mg-10Ni-2Mm alloy is favoured by melt spinning technique as larger amount of hydrogen was stored by the material prepared through this technique than that prepared using mechanical alloying. However, $YMg_{11}Ni$ in Table 5 showed a different trend. The rate constants of materials prepared by mechanical alloying are better than those of materials synthesized by melt spinning technique. This indicates that mechanical alloying plays a significant influence on increasing the hydrogenation kinetics of the alloys as compared to melt spinning.

Table 5. Comparison between kinetics and hydrogen content of as-milled and as-spun techniques.

Material	Rate constant, min^{-1}		Absorption Capacity, wt. %		Conditions	References
Alloy material	As-milled	As-spun	As-milled	As-spun		
Mg–10Ni–2Mm	0.04	0.01	3.21	4.22	573 K, 2.0 MPa	[92]
$YMg_{11}Ni$	0.05	0.03	4.72	4.07	593 K, 3.0 MPa	[44]

Hydrogenation of metal alloys is usually governed by three steps, namely (i) dissociation of hydrogen molecules into atomic hydrogen, (ii) the penetration of atomic hydrogen through grain boundaries, and (iii) the binding of atomic hydrogen on the metal, transforming into metal hydride [44]. The rate-determining step of the as-milled alloy is most likely ascribed to the ball milling rendering the second step, i.e., the penetration of atomic hydrogen through grain boundaries. The milling of as-cast alloys during mechanical alloying results in reduction of particle size and increment of high surface-to-volume ratio of alloy particles [93]. Due to these two changes, the distance for atomic hydrogen penetrating through the material is extremely reduced, which facilitates absorption greatly.

3.6. *Surface Modification of Ball Milled Metal Alloys*

Better and improved hydrogenation kinetics are achieved by nano-crystalline and/or amorphous metal alloys synthesized by mechanical alloying, but such materials still encounter several limitations, including impurity interactions, which include poisoning of the metal by impure gases such as carbon monoxide, sulphur and carbon dioxide; disproportionation during absorption/desorption cycling; retardation; innocuous and difficulty of initial activation [94]. These shortcomings result in rapid loss of hydrogen capacity with cycling, caused by impurities strongly or irreversibly absorbed on the surface active sites, and loss in absorption kinetics due to surface blanketing [95]. To overcome the above-mentioned problem, nano-structuring of ball milled materials and surface modification technology can be employed before hydrogen sorption can take place [96]. Surface modification involves the formation of a protective layer on the surface of metal alloy, thus preventing oxide-containing film to form. The most common surface modification technologies include fluorination, potassium borohydride, Ni-weak acid, ion implantation, hydrochloric (HCl) acid, platinum group metals (PGMs) decomposition and Y_3O_2 deposition. Yeung et al. [97] reported that the surface modification techniques PGMs deposition is the most attractive technology due to its ability to simultaneously prevent deactivation by guarding the catalytically active sites on the surface of bulk material against and enhance the hydriding kinetics during hydrogen absorption [97–100]. Moreover, this technique does not only enhance the hydriding rate, but also the absorption capacity of the alloys [99]. In particular, palladium is capable of catalyzing hydrogen sorption reactions, promoting easy activation, and improving poisoning tolerance. It has high hydrogen affinity and it is impermeable to larger molecules such as CO_2 and sulphur [82]. Figure 10a–c show different mechanisms of hydrogen dissociation on the surface of Pd-treated AB_5-type alloys [100]. During the hydrogen spillover mechanism hydrogen molecules are dissociated into hydrogen atoms on the surface of palladium, which then spillover to the active sites of the bulk alloy [98]. Pd thin film on the surface of AB_5 alloy dissociates hydrogen molecules and allows the resulting hydrogen atoms to pass through the film as they diffuse into the bulk alloy. However, this technique is costly as huge amount of Pd is utilized [99].

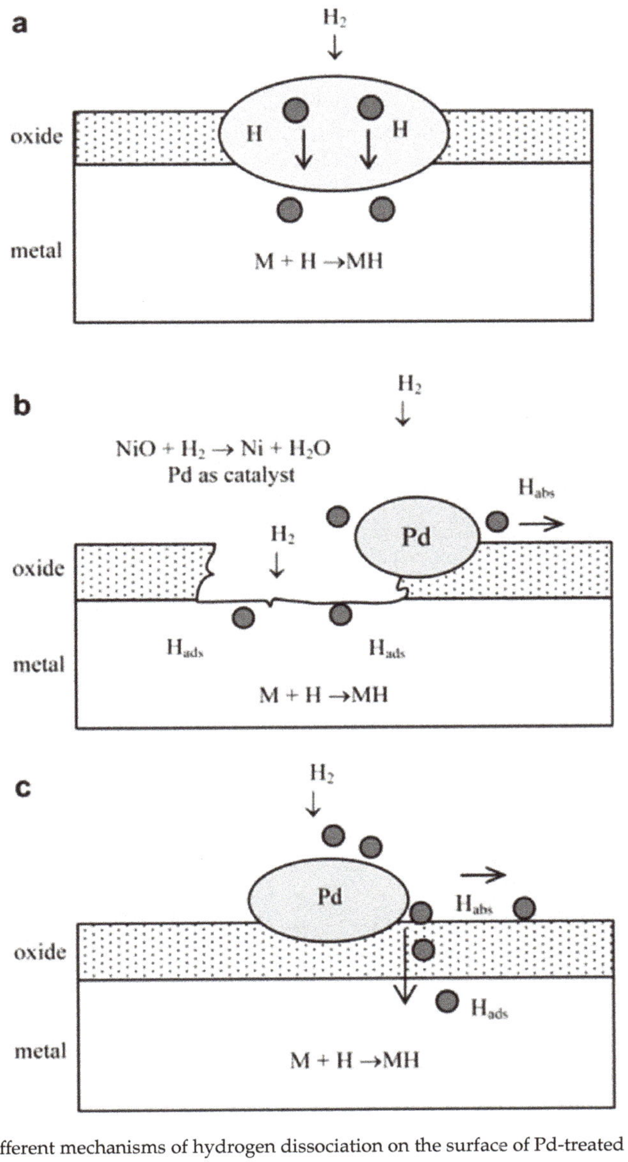

Figure 10. Different mechanisms of hydrogen dissociation on the surface of Pd-treated alloy material; (**a**) hydrogen diffusion through Pd window, (**b**) hydrogen spillover with oxide layer being reduced by spilt hydrogen from palladium, and (**c**) hydrogen spillover with hydrogen atoms diffusing through the oxide layer, reproduced from [100], with permission from Elsevier, 2009.

Moreover, PGM-treated materials also exhibit the power to absorb hydrogen at low temperature, without the activation step prior to absorption, and enhanced kinetics of hydrogen sorption processes [101]. This method is extremely active towards dissociative hydrogen chemisorption and permeable for hydrogen atoms; enabling hydrogen to rapidly absorb into the bulk material while still maintaining the hydrogenation activity even after exposure to the impurities inhibiting hydrogen sorption [102]. The properties being improved include surface catalytic activity, specific surface area layer, activation characteristics, as well as protective nature of the complexes against poisoning materials, such as sulphur, water vapour and carbon oxides [82].

Davids et al. [103] exploited hydrogenation behaviour of ball-milled TiFe alloy surface modified using a metal-organic chemical vapour deposition technique (MOCVD), by the thermal decomposition of palladium (II) acetylacetonate (Pd[acac]$_2$), mixed with the powder of the parent alloy. The authors reported that alloy materials containing 0.5% Pd have better hydrogenation behaviour than materials containing 1% Pd. The same trend in deterioration of hydrogenation performance, with an increase in Pd content, was observed by Pasquini and co-workers [104] who varied Pd content (corresponding to 4, 8 and 13 wt. %) on the surface of Mg metal. The observations are due to the theory that large amount of metal deposits on the surface of TiFe alloy and Mg metal results in slow penetration of hydrogen over a given period [104], as a result of fracture and defect blockage of the parent alloy material, which are usually formed during ball milling. Doyle et al. [105] recommended that the total weight of PGMs during surface modification of metal hydride-forming alloy should be in trace amounts (≤0.1wt. %) to avoid losses in hydrogenation behaviour. Another interesting study in literature was conducted by Modibane et al. [106] in which the authors carried out ball milling of AB$_5$-type materials out using a FRITSCH Pulverisette planetary mill and subsequently deposited Pd nanoparticles using autocatalytic Pd deposition technique. Upon ball milling of the parent alloy (Figure 11a), porous agglomerates of micro-size particles (Figure 11b,c) were obtained due to blending and grinding of parent particles. Porous particles are fancied by Parker et al. [35] for enhancement of absorption kinetics. Palladium nanoparticles, which showed discontinuous nature and near-spherical shape, partially cover the AB$_5$ materials (Figure 11d–f) and are responsible for dissociation of hydrogen molecules [106].

Table 6 reveals some of reported literature on surface modified metal alloys synthesized through mechanical alloying. Surface modification techniques herein are based on PGMs deposition and co-deposition. All surface modified alloys exhibit better hydrogenation kinetics as compared to their as-milled counterparts. Variations arise when comparing the hydrogen capacities. For instance, Pt deposition on as-milled $La_{0.9}Pr_{0.05}Nd_{0.05}Al_{0.3}Mn_{0.4}Co_{0.65}Ni_{3.5}$ alloy results in increase of capacity, while Ru deposition on the same alloy lowers the hydrogen capacity. This is attributed to poor or little hydrogen affinity associated with Ru.

Table 6. Hydrogenation properties of surface modified metal alloys prepared by mechanical alloying.

Alloy Type	Metal Alloy	Rate Constant, min^{-1}	Hydrogen Capacity, wt. %	Hydrogenation Conditions	References
As-milled	$LaNi_5$	0.04	5.30	313 K, 1.5 MPa	[28]
Pd deposition	Pd-$LaNi_5$	2.38	5.18	313 K, 1.5 MPa	
As-milled	Mg_2Ni	0.06	3.55	673 K, 2.0 MPa	[28]
Pd deposition	Pd-Mg_2Ni	0.59	2.64	573 K, 2.0 MPa	
As-milled	$LaNi_{4.25}Al_{0.75}$	4.8×10^{-1}	3.21	293 K, 1.5 MPa	[107]
Pd deposition	Pd-$LaNi_{4.25}Al_{0.75}$	1.33	3.89	293 K, 1.5 MPa	
As-milled	$La_{0.9}Pr_{0.05}Nd_{0.05}Al_{0.3}Mn_{0.4}Co_{0.65}Ni_{3.5}$	2.4×10^{-3}	0.45	293 K, 0.5 MPa	[108]
Pt deposition	Pt-$La_{0.9}Pr_{0.05}Nd_{0.05}Al_{0.3}Mn_{0.4}Co_{0.65}Ni_{3.5}$	4.5×10^{-2}	0.83	298 K, 0.1 MPa	
Ru deposition	Ru-$La_{0.9}Pr_{0.05}Nd_{0.05}Al_{0.3}Mn_{0.4}Co_{0.65}Ni_{3.5}$	3.3×10^{-2}	0.34	298 K, 0.1 MPa	
Pt-Ru co-deposition	Pt-Ru-$La_{0.9}Pr_{0.05}Nd_{0.05}Al_{0.3}Mn_{0.4}Co_{0.65}Ni_{3.5}$	2.9×10^{-2}	0.79	298 K, 0.1 MPa	
As-milled	$La_{0.40}Ce_{0.48}(Nd,Pr)_{0.16}Ni_{3.34}Co_{0.64}Al_{0.63}Mn_{0.58}$	1.6×10^{-4}	1.24	293 K, 0.5 MPa	[101]
Pd deposition	Pd-$La_{0.40}Ce_{0.48}(Nd,Pr)_{0.16}Ni_{3.34}Co_{0.64}Al_{0.63}Mn_{0.58}$	6.2×10^{-4}	2.75	293 K, 0.5 MPa	
Pd-Ni co-deposition	Pd-Ni-$La_{0.40}Ce_{0.48}(Nd,Pr)_{0.16}Ni_{3.34}Co_{0.64}Al_{0.63}Mn_{0.58}$	1.1×10^{-3}	3.60	293 K, 0.5 MPa	

Figure 11. Low-((**a**,**b**), Hitachi X-650 EM) and high-resolution ((**c**–**f**), Zeiss Ultra) SEM images of raw AB5 alloy (**a**), the alloy ball milled without Pd black (**b**), and the alloy ball milled with Pd black after autocatalytic Pd deposition at different magnifications (**c**–**f**), reproduced from [106], with permission from Elsevier, 2018.

4. Conclusions

Hydrogenation kinetics and storage in solid state materials is a major concern in developing an up-and-coming hydrogen technology and economy. The present work reviews some challenges regarding the hydride-forming alloys, particularly their hydriding kinetics and capacity. In this regard, developments on hydrogenation performances of metal alloys, prepared by the mechanical alloying technique, were discussed. Moreover, this technique was compared with other preparation techniques, such as spark plasma sintering and melt spinning. The work is carried out in an attempt to facilitate suitable alloy preparation technique that can extensively improve the hydrogenation kinetics of different alloy materials. Although, much attention has been given to the techniques that are discussed in this work, with the aim of minimizing dehydrogenation temperature, enhancing the kinetics and poisoning-tolerance of different alloys, further research is needed. The collected literature data reveal encouraging improvements in terms of hydrogen storage kinetics through mechanical alloying. In the light of the achievements, through the combination of mechanical alloying and melt spinning techniques there is high potential in developing better metal hydride materials with high hydriding kinetics. Mechanically alloyed materials still suffer from cycling stability; they are prone to oxide layers formation, which hinders rapid absorption of hydrogen; and they also exhibit poisonous-intolerance. As a result, we further reviewed different surface modification techniques that can overcome these drawbacks. Surface modification of mechanically alloyed materials was overlooked at and it was seen that it promotes hydrogen storage kinetics by forming a protective layer on the surface of bulk alloy, thereby preventing the formation of oxide layer. Moreover, surface modification through deposition of PGMs on the surface of alloy materials stimulates poisonous tolerance of parent alloys.

Author Contributions: Conceptualization, K.D.M., and M.V.L.; funding acquisition, K.D.M. and M.J.H.; project administration, K.D.M.; Supervision, K.D.M., and M.V.L.; writing—original draft, T.R.S., T.C.M., M.W.D., K.D.M., M.J.H. and M.V.L. All authors have read and agreed to the published version of the manuscript.

Funding: This research was supported by the National Research Foundation under Thuthuka program (UIDs. 117727 and 118113), University of Limpopo (Research Development Grants R202 and R232) and Sasol Inzalo Foundation, South Africa.

Conflicts of Interest: The authors declare no conflict of interest.

References

1. Andreas, R. Hydrogen storage methods. *Naturwissenschaften* **2004**, *91*, 157–172.
2. Schlapbach, L.; Züttel, A. Hydrogen-storage materials for mobile applications. *Mater. Sustain. Energy* **2011**, *414*, 265–270.
3. Berlouis, L.E.A.; Cabrera, E.; Hall, P.J. Thermal analysis investigation of hydriding properties of nanocrystalline Mg–Ni–and Mg–Fe-based alloys prepared by high-energy ball milling. *J. Alloys Compd.* **2000**, *305*, 82–89. [CrossRef]
4. Sakintuna, B.; Lamari-darkrim, F.; Hirscher, M. Metal hydride materials for solid hydrogen storage: A review. *Int. J. Hydrog. Energy* **2007**, *32*, 1121–1140. [CrossRef]
5. Crivello, J.C.; Dam, B.; Yartys, V.A. Review of magnesium hydride-based materials: Development and optimisation. *Appl. Phys. A Mater. Sci. Process.* **2016**, *122*, 1–20. [CrossRef]
6. Liu, T.; Chen, C.; Qin, C.; Li, X. Improved hydrogen storage properties of Mg-based nanocomposite by addition of $LaNi_5$ nanoparticles. *Int. J. Hydrogen Energy* **2014**, *39*, 18273–18279. [CrossRef]
7. Yartys, V.A.; Lototskyy, M.V. An overview of hydrogen storage methods. In *Hydrogen Materials Science and Chemistry of Carbon Nanomaterials*; Springer: Dordrecht, The Netherland, 2004; pp. 75–104.
8. Berube, V.; Chen, G.; Dresselhaus, M.S. Impact of nanostructuring on the enthalpy of formation of metal hydrides. *Int. J. Hydrog. Energy* **2008**, *33*, 4122–4131. [CrossRef]
9. Pandey, S.K.; Srivastava, A.; Srivastava, O.N. Improvement in hydrogen storage capacity in $LaNi_5$ through substitution of Ni by Fe. *Int. J. Hydrog. Energy* **2007**, *32*, 2461–2465. [CrossRef]
10. Kazakov, A.N.; Dunikov, D.O.; Mitrokhin, S.V. AB_5-type intermetallic compounds for biohydrogen purification and storage. *Int. J. Hydrog. Energy* **2016**, *41*, 21774–21779. [CrossRef]
11. Takeda, H.; Kabutomori, T.; Ohnishi, K. Metal hydride air-conditioning. *Encycl. Life Support Syst.* **2009**, *2*, 249–263.
12. Davids, W.; Lototskyy, M.V.; Linkov, V. Chemical surface modification for the improvement of the hydrogenation kinetics and poisoning resistance of TiFe. *J. Alloys Compd.* **2011**, *509S*, S770–S774.
13. Semboshi, S.; Masahashi, N.; Hanada, S. Degradation of hydrogen absorbing capacity in cyclically hydrogenated $TiMn_2$. *Acta Metall. Sin.* **2001**, *49*, 927–935. [CrossRef]

14. Andreasen, A. Predicting formation enthalpies of metal hydrides. *Nature* **2001**, *414*, 353–358.
15. Chebab, S.; Abdellaoui, M.; Latroche, M.; Paul-boncour, V. $LaCaMgNi_9$ synthesized by mechanical alloying: Structural and electrochemical characterization. *J. Tunis. Chem. Soc.* **2016**, *18*, 52–59.
16. Young, K. Increase in the Surface Catalytic Ability by Addition of Palladium in C14 Metal Hydride Alloy. *Batteries.* **2017**, *3*, 26. [CrossRef]
17. Zhang, Y.; Li, B.; Ren, H.; Ding, X.; Liu, X.; Chen, L. An investigation on hydrogen storage kinetics of nanocrystalline and amorphous $Mg_2Ni_{1-x}Co_x(x = 0 - 0.4)$ alloy prepared by melt spinning. *J. Alloys Compd.* **2011**, *509*, 2808–2814. [CrossRef]
18. Simchi, H.; Kaflou, A.; Simchi, A. Synergetic effect of Ni and Nb_2O_5 on dehydrogenation properties of nanostructured MgH_2 synthesized by high-energy mechanical alloying. *Int. J. Hydrog. Energy* **2009**, *34*, 7724–7730. [CrossRef]
19. Suryanarayana, C. Mechanical alloying and milling. *Prog. Mater. Sci.* **2001**, *46*, 1–184. [CrossRef]
20. Suñol, J.J.; Fort, J. Materials developed by mechanical alloying and melt spinning. *Int. Rev. Phys.* **2008**, *2*, 31–35.
21. Scudino, S.; Sakaliyska, M.; Surreddi, K.B. Mechanical alloying and milling of Al-Mg alloys. *J. Alloys Compd.* **2009**, *483*, 2–7. [CrossRef]
22. Poondi, D.; Singh, J. Synthesis of metastable silver-nickel alloys by a novel laser-liquid-solid interaction technique. *J. Mater. Res.* **2000**, *35*, 2467–2476.
23. Xing, N.; Wu, Y.; Zhou, S. Improved hydrogenation-dehydrogenation characteristics of nanostructured melt-spun Mg-10Ni-2Mm alloy processed by rapid solidification. *Prog. Nat. Sci. Mater. Int.* **2010**, *20*, 49–53. [CrossRef]
24. Ouyang, L.Z.; Qin, F.X.; Zhu, M. The hydrogen storage behavior of Mg_3La and $Mg_3LaNi_{0.1}$. *Scr. Mater.* **2006**, *55*, 1075–1078. [CrossRef]
25. Ares, J.R.; Cuevas, F. Influence of thermal annealing on the hydrogenation properties of mechanically milled AB_5-type alloys. *Mater. Sci. Eng.* **2004**, *108*, 76–80. [CrossRef]
26. Guevara, L.; Welsh, R.; Atwater, M.A. Parametric Effects of Mechanical Alloying on Carbon Nanofiber Catalyst Production in the Ni-Cu System. *Metals (Basel)* **2018**, *8*, 286. [CrossRef]
27. Shang, C.X.; Bououdina, M.; Guo, Z.X. Mechanical alloying and electronic simulations of (MgH_2 + M) systems (M = Al, Ti, Fe, Ni, Cu and Nb) for hydrogen storage. *Int. J. Hydrog. Energy* **2004**, *29*, 73–80. [CrossRef]
28. Zaluski, L.; Zaluska, A.; Strrm-olsen, J.O.; Schulz, R. Catalytic effect of Pd on hydrogen absorption in mechanically alloyed Mg_2Ni, $LaNi_5$ and FeTi. *J. Alloys Compd.* **1995**, *217*, 295–300. [CrossRef]
29. Iturbe-garcía, J.L.; García-núñez, M.R.; López-muñoz, B.E. Synthesis of the Mg_2Ni Alloy Prepared by Mechanical Alloying Using a High Energy Ball Mill. *J. Mex. Chem. Soc.* **2010**, *54*, 46–50. [CrossRef]
30. Kumar, S.; Tiwari, G.P.; Krishnamurthy, N. High performance FeTi-3.1 mass% V alloy for on board hydrogen storage solution. *Energy* **2014**, *75*, 520–524. [CrossRef]
31. Andreasen, A. Hydrogenation properties of Mg–Al alloys. *Int. J. Hydrog. Energy* **2008**, *33*, 7489–7497. [CrossRef]
32. Gleiter, H. Nanostrctured materials: Basic concepts and microstructure. *Acta Mater.* **2000**, *48*, 1–29. [CrossRef]
33. Friedrichs, O.; Kolodziejczyk, L.; Fernandez, A. Synthesis of nanocrystalline MgH_2 powder by gas-phase condensation and in situ hydridation: TEM, XPS and XRD study. *J. Alloys Compd.* **2007**, *435*, 721–724. [CrossRef]
34. Lopez-Suarez, A. Effect of absorption and desorption of hydrogen in Ti and Ti alloys. In *New Advances in Hydrogenation Processes: Fundamental and Applications*; Iran Polymer and Petrochemical Institute: Teheran, Iran, 2017. [CrossRef]
35. Parker, S.F.; Konrad, M.; Wieland, S.D. The effect of particle size, morphology and support commercial catalysts. *Chem. Sci.* **2019**, *10*, 480–489. [CrossRef] [PubMed]
36. Ismail, M.; Mustafa, N.S.; Yap, A.H. Catalytic effect of $CeCl_3$ on the hydrogen storage properties of MgH_2. *Mater. Chem. Phys.* **2016**, *170*, 77–82. [CrossRef]
37. Lototskyy, M.; Davids, M.W.; Pollet, B.G. Magnesium-based hydrogen storage nanomaterials prepared by high energy reactive ball milling in hydrogen at the presence of mixed titanium—iron oxide. *J. Alloys Compd.* **2015**, *645*, S454–S459. [CrossRef]

38. Bouaricha, S.; Dodelet, J.P.; Schulz, R. Hydriding behavior of Mg–Al and leached Mg–Al compounds prepared by high-energy ball-milling. *J. Alloys Compd.* **2000**, *297*, 282–293. [CrossRef]
39. Liang, G.; Huot, J.; Boily, S.; Van Neste, A.; Schulz, R. Hydrogen storage properties of the mechanically milled MgH $_2$–V nanocomposite. *J. Alloys Compd.* **1999**, *291*, 295–299. [CrossRef]
40. Huang, H.X.; Huang, K.L.; Liu, S.Q. Microstructures and electrochemical properties of $Mg_{0.9}Ti_{0.1}Ni_{1-x}M_x$ (M = Co, Mn; x = 0, 0.1, 0.2) hydrogen storage alloys. *Powder Technol.* **2010**, *198*, 144–148. [CrossRef]
41. Vermeulen, P.; Niessen, R.A.H.; Notten, P.H.L. Hydrogen storage in metastable $Mg_yTi_{(1-y)}$ thin films. *Electrochem. Commun.* **2006**, *8*, 27–32. [CrossRef]
42. Luo, W. ($LiNH_2$-MgH_2): A viable hydrogen storage system. *J. Alloys Compd.* **2004**, *381*, 284–287. [CrossRef]
43. Castro, F.J.; Fuster, V.; Urretavizcaya, G. Hydrogen sorption properties of a MgH_2-10wt. % graphite mixture. *J. Alloys Compd.* **2011**, *509*, S595–S598. [CrossRef]
44. Zhang, Y.; Zhang, W.; Qi, W.B.Y. A Comparison Study of Hydrogen Storage Thermodynamics and Kinetics of YMg11Ni Alloy Prepared by Melt Spinning and Ball Milling. *Acta Metall. Sin.* **2017**, *30*, 1040–1048. [CrossRef]
45. Imamura, H.; Hashimoto, Y.; Sakata, Y. Preparation and Properties of Ball-Milled MgH_2/Al Nanocomposites for Hydrogen Storage. *Mater. Trans.* **2014**, *55*, 572–576. [CrossRef]
46. *CRYSTMET Database, version 5.0.0*; Toth Information Systems Inc.: Ottawa, ONT, Canada, 2013.
47. Lototskyy, M.; Wafeeq, M.; Yartys, V.A. Nanostructured hydrogen storage materials prepared by high-energy reactive ball milling of magnesium and ferrovanadium. *Int. J. Hydrog. Energy* **2019**, *44*, 6687–6701. [CrossRef]
48. Williams, M.; Lototskyy, M.V.; Pollet, B.G. Hydrogen absorption study of high-energy reactive ball milled Mg composites with palladium additives. *J. Alloys Compd.* **2013**, *580*, s144–s148. [CrossRef]
49. Bhatnagar, A.; Pandey, S.K.; Srivastava, O.N. Catalytic effect of carbon nanostructures on the hydrogen storage properties of MgH_2-$NaAlH_4$ composite. *Int. J. Hydrog. Energy* **2014**, *39*, 14240–14246. [CrossRef]
50. Yahya, M.S.; Ismail, M. Catalytic effect of $SrTiO_3$ on the hydrogen storage behaviour of MgH_2. *J. Energy Chem.* **2019**, *28*, 46–53. [CrossRef]
51. Hanada, N.; Ichikawa, T.; Fujii, H. Remarkable improvement of hydrogen sorption kinetics in magnesium catalyzed with Nb_2O_5. *J. Alloys Compd.* **2006**, *420*, 46–49. [CrossRef]
52. Juahir, N.; Mustafa, N.S.; Sinin, A.M. Improved hydrogen storage properties of MgH_2 by addition of Co_2NiO nanoparticles. *RSC Adv.* **2015**, *5*, 60983–60989. [CrossRef]
53. Malka, I.E.; Pisarek, M.; Bystrzycki, J. A study of the ZrF_4, NbF_5, TaF_5, and $TiCl_3$ influences on the MgH2 sorption properties. *Int. J. Hydrog. Energy* **2011**, *36*, 12909–12917. [CrossRef]
54. Wu, C.Z.; Wang, P.; Cheng, H.M. Hydrogen storage properties of MgH_2/SWNT composite prepared by ball milling. *J. Alloys Compd.* **2006**, *420*, 278–282. [CrossRef]
55. El-eskandarany, M.S.; Shaban, E.; Al-duweesh, A. Superior catalytic effect of nanocrystalline big-cube Zr_2 Ni metastable phase for improving the hydrogen sorption / desorption kinetics and cyclability of MgH_2 powders. *Energy* **2015**, *91*, 274–282. [CrossRef]
56. Santos, S.F.; Ishikawa, T.T.; Huot, J. MgH_2 + FeNb nanocomposites for hydrogen storage. *Mater. Chem. Phys.* **2014**, *147*, 557–562. [CrossRef]
57. Liang, G.; Huot, J.; Boily, S.; Van Neste, A.; Schulz, R. Hydrogen storage in mechanically milled Mg–$LaNi_5$ and MgH_2–$LaNi_5$ composites. *J. Alloys Compd.* **2000**, *297*, 261–265. [CrossRef]
58. Chen, X.Q.; Podloucky, R.; Rogl, P. Computational and experimental study of phase stability, cohesive properties, magnetism and electronic structure of TiMn2. *Acta Mater.* **2003**, *51*, 1239–1247. [CrossRef]
59. Erika, T.; Sebastian, C.; Fernando, Z. Temperature performance of AB5 hydrogen storage alloy for Ni-MH batteries. *Int. J. Hydrog. Energy* **2016**, *41*, 19684–19690. [CrossRef]
60. Hayakawa, H.; Akiba, E.; Kohno, T. Crystal Structures of La–Mg–Nix (x = 3-4) System Hydrogen Storage Alloys. *Mater. Trans.* **2005**, *46*, 1393–1401. [CrossRef]
61. Boeije, M.F.J.; Delczeg-czirjak, E.K.; Brück, E. On the phase stability of CaCu 5 -type compounds. *J. Alloys Compd.* **2017**, *722*, 549–554. [CrossRef]
62. Singh, A.; Singh, B.K.; Srivastava, O.N. Studies on improvement of hydrogen storage capacity of AB5 type: MmNi4.6Fe0.4 alloy. *Int. J. Hydrog. Energy* **2004**, *29*, 1151–1156.
63. Roberts, M.; Smith, R.I.; Hull, S. In situ investigation of commercial Ni(OH)2 and LaNi5-based electrodes by neutron powder diffraction. *J. Mater. Res.* **2015**, *30*, 407–416.
64. Cermak, J.; David, B. Catalytic effect of Ni, Mg2Ni and Mg2NiH4 upon hydrogen desorption from MgH2. *Int. J. Hydrog. Energy* **2011**, *6*, 3–9. [CrossRef]

65. Khajavi, S.; Rajabi, M.; Huot, J. Effect of cold rolling and ball milling on first hydrogenation of Ti0.5Zr0.5 (Mn1-xFex) Cr1, x = 0, 0.2, 0.4. *J. Alloys Compd.* **2019**, *775*, 912–920. [CrossRef]
66. Gkanas, E.I.; Khzouz, M.; Makridis, S.S. Synthesis and Hydrogen Sorption Characteristics of Mechanically Alloyed Mg(NixMn1-x)2 Intermetallics. *Mater. Sci.* **2017**, *12*, 257–268.
67. Srivastava, S.; Panwar, K. Effect of transition metals on ball-milled $MmNi_5$ hydrogen storage alloy. *Mater. Renew. Sustain. Energy* **2015**, *4*, 19. [CrossRef]
68. Broz, M.E.; Cook, R.F.; Whitney, D.L. Microhardness, toughness, and modulus of Mohs scale minerals. *Am. Mineral.* **2006**, *91*, 2006. [CrossRef]
69. Chen, Y.; Sequeira, C.A.C.; Song, X.; Neto, R.; Wang, Q. Polytypism of La–Ni phases in multicomponent AB_5 type hydride electrode alloys. *Int. J. Hydrog. Energy* **2002**, *27*, 63–68. [CrossRef]
70. Dobrovolsky, V.D. The correlation between ionicity of metal-hydrogen bonds in hydrides and their thermal firmness. In *Hydrogen Materials Science and Chemistry of Carbon Nanomaterials*; Springer: Dordrecht, The Netherlands, 2007; pp. 421–428.
71. Joseph, B.; Schiavo, B. Effects of ball-milling on the hydrogen sorption properties of $LaNi_5$. *J. Alloys Compd.* **2009**, *480*, 912–916. [CrossRef]
72. Koultoukis, E.D.; Christodoulou, C.N.; Fruchart, D. High-Temperature Activated AB_2 Nanopowders for Metal Hydride Hydrogen Compression. *Int. J. energy Res.* **2014**, *38*, 477–486. [CrossRef]
73. Ouyang, L.; Huang, J.; Zhu, M. Progress of hydrogen storage alloys for Ni-MH rechargeable power batteries in electric vehicles: A review. *Mater. Chem. Phys.* **2017**, *200*, 164–178. [CrossRef]
74. Stein, F.; Palm, M.; Sauthoff, G. Structure and stability of Laves phases. Part I. Critical assessment of factors controlling Laves phase stability. *Intermetallics* **2004**, *12*, 713–720. [CrossRef]
75. Hammerschmidt, T.; Ladines, A.N.; Drautz, R. Crystal-Structure Analysis with Moments of the Density-of-States: Application to Intermetallic Topologically Close-Packed Phases. *Crystals* **2016**, *6*, 18. [CrossRef]
76. Ioannidou, A.; Makridis, S.; Kikkinides, E.S. Structural and Hydrogenation Properties of $Zr_{0.9}Ti_{0.1}Cr_{1.2-x}V_{0.8}Ni_x$ (x=0, 0.4) Compounds. *Mater. Sci. Forum.* **2010**, *636*, 26–31.
77. Liu, P.; Xie, X.; Liu, T. Hydrogen storage properties of $(Ti_{0.85}Zr_{0.15})_{1.05}Mn_{1.2}Cr_{0.6}V_{0.1}M_{0.1}$ (M=Ni, Fe, Cu) alloys easily activated at room temperature. *Prog. Nat. Sci. Mater. Int.* **2017**, *27*, 652–657. [CrossRef]
78. Zhang, Y.; Xu, S.; Zhao, D. Hydrogen storage kinetics of nanocrystalline and amorphous Cu–Nd-added Mg_2Ni-type alloys. *Trans. Nonferrous Met. Soc. China* **2014**, *24*, 3524–3533. [CrossRef]
79. Liu, J.; Song, X.P.; Chen, G.L. Hydrogen storage performance of Mg-based composites prepared by spark plasma sintering. *J. Alloys Compd.* **2009**, *486*, 338–342. [CrossRef]
80. Becker, H. Processing of bulk Al7075 alloy by spark plasma sintering. *Mater. Sci. Eng.* **2017**, *179*, 2050.
81. Pei, P.; Song, X.; Liu, J.; Song, A.; Zhang, P.; Chen, G. Study on the hydrogen desorption mechanism of a Mg-V composite prepared by SPS. *Int. J. Hydrog. Energy* **2011**, *37*, 984–989. [CrossRef]
82. Huang, H.; Huang, K. Effect of Fluorination Treatment on Electrochemical Properties of $M_1Ni_{3.5}Co_{0.6}Mn_{0.4}Al_{0.5}$ Hydrogen Storage Alloy. *J. Braz. Chem. Soc.* **2012**, *23*, 951–957.
83. Schwars, R.B.; Johnson, W.L. Formation of an amorphous alloy by solid-state reaction of the pure polycrystalline metals. *Phys. Rev. Lett.* **1983**, *51*, 415. [CrossRef]
84. Termsuksawad, P.; Niyomsoan, S.; Gavra, Z. Measurement of hydrogen in alloys by magnetic and electronic techniques. *J. Alloys Compd.* **2004**, *373*, 86–95. [CrossRef]
85. Zadorozhnyy, V.Y.; Milovzorov, G.S.; Kaloshkin, S.D. Preparation and hydrogen storage properties of nanocrystalline TiFe synthesized by mechanical alloying. *Prog. Nat. Sci. Mater. Int.* **2017**, *27*, 149–155. [CrossRef]
86. Møller, K.T.; Sheppard, D.; Ravnsbæk, D.B. Complex Metal Hydrides for Hydrogen, Thermal and Electrochemical Energy Storage. *Energies* **2017**, *10*, 1645. [CrossRef]
87. Song, X.; Zhang, P.; Chen, G. The role of spark plasma sintering on the improvement of hydrogen storage properties of Mg-based composites. *Int. J. Hydrog. Energy* **2010**, *35*, 8080–8087. [CrossRef]
88. Pei, P.; Song, X.P.; Chen, G.L. Improving hydrogen storage properties of Laves phase related BCC solid solution alloy by SPS preparation method. *Int. J. Hydrog. Energy* **2009**, *34*, 8597–8602. [CrossRef]
89. Lv, P.; Guzik, M.N.; Huot, J. Effect of ball milling and cryomilling on the microstructure and first hydrogenation properties of TiFe + 4 wt.% Zr alloy. *J. Mater. Res. Technol.* **2019**, *10*, 1016. [CrossRef]

90. Zhang, Q.A.; Zhang, L.X.; Wang, Q.Q. Crystallization behavior and hydrogen storage kinetics of amorphous $Mg_{11}Y_2Ni_2$ alloy. *J. Alloys Compd.* **2013**, *551*, 376–381. [CrossRef]
91. Zhang, D.Z.Y.; Li, B.; Ren, H.; Li, X.; Qi, T. Enhanced Hydrogen Storage Kinetics of Nanocrystalline and Amorphous Mg_2Ni-type Alloy by Melt Spinning. *Materials (Basel)* **2011**, *4*, 274–287. [CrossRef]
92. Wu, Y.; Lototsky, M.V.; Yartys, V.A. Microstructure and hydrogenation behavior of ball-milled and melt-spun Mg–10Ni–2Mm alloys. *J. Alloys Compd.* **2008**, *466*, 176–181. [CrossRef]
93. Lass, E.A. Hydrogen storage measurements in novel Mg-based nanostructured alloys produced via rapid solidification and devitrification. *Int. J. Hydrog. Energy* **2011**, *36*, 10787–10796. [CrossRef]
94. Modibane, K.D.; Williams, M.; Lototskyy, M. Poisoning-tolerant metal hydride materials and their application for hydrogen separation from CO_2/CO containing gas mixtures. *Int. J. Hydrog. Energy* **2013**, *38*, 9800–9810. [CrossRef]
95. Aguey-Zinsou, K.F.; Ares Fernandez, J.R.; Klassen, T. Effect of Nb_2O_5 on MgH_2 properties during mechanical milling. *Int. J. Hydrog. Energy* **2007**, *32*, 221–233.
96. Williams, M.; Nechaev, A.N.; Lototskyy, M.V. Influence of aminosilane surface functionalization of rare earth hydride-forming alloys on palladium treatment by electroless deposition and hydrogen sorption kinetics of composite materials. *Mater. Chem. Phys.* **2009**, *115*, 136–141. [CrossRef]
97. Yeung, K.L.; Christiansen, S.C.; Varma, A. Palladium composite membranes by electroless plating technique Relationships between plating kinetics, film microstructure and membrane performance. *J. Memb. Sci.* **1999**, *159*, 107–122. [CrossRef]
98. Parambhath, V.B.; Nagar, R.; Ramaprabhu, S. Effect of Nitrogen Doping on Hydrogen Storage Capacity of Palladium Decorated Graphene. *Langmuir* **2012**, *28*, 7826–7833. [CrossRef]
99. Charbonnier, M.; Romand, M.; Goepfert, Y. Palladium reduction: A key step for the electroless Ni metallization of insulating substrates by a tin-free process. *Thin Solid Film.* **2006**, *515*, 1623–1633. [CrossRef]
100. Shan, X.; Payer, J.H.; Jennings, W.D. Mechanism of increased performance and durability of Pd-treated metal hydriding alloys. *Int. J. Hydrog. Energy* **2009**, *34*, 363–369. [CrossRef]
101. Denys, R.V.; Lototskyy, M.V.; Linkov, V.M.; Williams, M. Palladium mixed-metal surface-modified AB_5-type intermetallides enhance hydrogen sorption kinetics. *S. Afr. J. Sci.* **2010**, *106*, 1–6.
102. Lototskyy, M.V.; Williams, M.; Yartys, V.A. Surface-modified advanced hydrogen storage alloys for hydrogen separation and purification. *J. Alloys Compd.* **2011**, *509*, 555–561. [CrossRef]
103. Davids, W.; Lototskyy, M.V.; Williams, M. Surface modification of TiFe hydrogen storage alloy by metal-organic chemical vapour deposition of palladium. *Int. J. Hydrog. Energy* **2011**, *36*, 9743–9750. [CrossRef]
104. Pasquini, L.; Callini, E.; Maurizio, C. Magnesium nanoparticles with transition metal decoration for hydrogen storage. *J. Nanoparticle Res.* **2011**, *13*, 5727–5737. [CrossRef]
105. Mark, I.; Doyle, L.; Benjamin, D. Hydrogen storage materials. *United States Pat.* **2000**, *6*, 165.
106. Modibane, K.D.; Lototskyy, M.; Davids, M.W.; Williams, M.; Hato, M.J.; Molapo, K.M. Influence of co-milling with palladium black on hydrogen sorption performance and poisoning tolerance of surface modified AB5-type hydrogen storage alloy. *J. Alloys Compd.* **2018**, *750*, 523–529. [CrossRef]
107. Ceramics, M.; Skorokhod, V.V.; Klimenko, V.P. Reversable hydriding of $LaNi_{5-x}Al_x$–Pd composite in the presence of carbon monoxide. *Powder Metall. Met. Ceram.* **2001**, *39*, 575–583.
108. Willey, D.B.; Pederzolli, D.; Harris, I.R. Low temperature hydrogenation properties of platinum group metal treated, nickel metal hydride electrode alloy. *J. Alloys Compd.* **2002**, *332*, 806–809. [CrossRef]

metals

MDPI

Article

A Novel Non-Equiatomic $(W_{35}Ta_{35}Mo_{15}Nb_{15})_{95}Ni_5$ Refractory High Entropy Alloy with High Density Fabricated by Powder Metallurgical Process

Bohua Duan, Yingrui Yu, Xinli Liu *, Dezhi Wang and Zhuangzhi Wu

School of Materials Science and Engineering, Central South University, Changsha 410083, China; dbh72@163.com (B.D.); yyr021189@163.com (Y.Y.); dzwang@csu.edu.cn (D.W.); zwu2012@csu.edu.cn (Z.W.)
* Correspondence: liuxinli@csu.edu.cn

Received: 10 October 2020; Accepted: 25 October 2020; Published: 29 October 2020

Abstract: A non-equiatomic refractory high entropy alloy (RHEA), $(W_{35}Ta_{35}Mo_{15}Nb_{15})_{95}Ni_5$ with high density of 14.55 g/cm^3 was fabricated by powder metallurgical process of mechanical alloying (MA) and spark plasma sintering (SPS). The mechanical alloying behavior of the metallic powders was studied systematically, and the microstructure and phase composition for both the powders and alloys were analyzed. Results show that the crystal consists of the primary solid solution and marginal oxide inclusion ($Nb_{5.7}Ni_4Ta_{2.3}O_2$). In addition, the maximum strength, yield strength and fracture strain are, 2562 MPa, 2128 MPa, 8.16%, respectively.

Keywords: refractory; high entropy alloy; mechanical alloying; phase transformation; mechanical properties

1. Introduction

With the rapid development of advanced science and technology in the fields of armor piercing projectile, counter weight balance, radiation shields, shaped charge liner in ordnance industry and high-density gradient materials in dynamic high-pressure physics, the demand for materials with high density has boosted wide study on heavy metals and alloys [1]. Recently, a novel type of advanced refractory metallic alloys, refractory high entropy alloys, have been regarded as a potential candidate. High entropy alloy is a kind of alloy with high entropy of more than 1.61 R (when the alloys have 5 components). The different atoms with different bonding energy and crystal structure make the crystal lattice complex, and solute phases lead to severe lattice distortion in high entropy alloy (HEAs). These distortions are proposed to be more severe than in conventional alloys. The diffusion of atoms in HEAs mainly occurs via the vacancy diffusion mechanism. Due to the different melting points of the components, the element with high migration rate will preferentially diffuse to the vacancies. However, different atoms in HEAs have different bonding energies. Once atoms fill in vacancy, the energy of the system will decrease or increase. As a result, compared to the conventional alloys, the atomic diffusion and phase transformation processes in HEAs are relatively slow. This is called the sluggish diffusion effect in HEAs [2]. Refractory high entropy alloys (RHEAs) possess many excellent properties, such as high strength and hardness, and excellent creep resistance, due to its lattice distortion effect and sluggish diffusion effect [2]. Unlike conventional alloys with one or two major elements, RHEAs are generally made up of multiple principle refractory elements (≥5), such as Cr, W, Ta, Mo, Nb, Ti, V, Zr, Hf, and other elements of Al, Ni, Co, Si, C and N [3] in equimolar or near-equimolar proportions (5–35 at%). For the moment, the HEAs was recognized consisting of at least four major metallic elements in non-equimolar ratios (5–35 at%). The most common RHEAs include WTaMoNb, HfNbTaTiZr, NbTiVZr, AlNbTiV and CrMoVW, which have received extensive attention due to the easy way to tailor the phase stability and phase microstructure to enhance mechanical properties [4].

Up to now, various fabrication methods have been adopted to manufacture the HEAs, such as vacuum arc-melting and casting, sintering, magnetron sputtering and laser cladding. Among these methods, vacuum arc-melting and casting is generally applied to product bulk HEAs. However, the defects such as segregation and inhomogeneous dendritic microstructures generally exist in HEAs which are prepared by casting, and may be harmful to mechanical properties [5]. A few HEA systems were manufactured by mechanical alloying (MA) process [6]. HEA powders produced by MA process exhibit many advantages, such as more homogeneous morphologies and precise composition control. However, the milling time which achieves alloying is overwhelmingly long [6,7]. Some researches described appropriate addition of Ni can reduce reaction energy barrier and significantly accelerate the reaction among brittle phase during MA process [8]. Spark plasma sintering (SPS) technique is a new technology for preparing materials, which has the distinctive characteristics of fast heating speed, short sintering time, controllable structure, energy saving and environmental protection, and can be used to prepare metals materials, ceramic materials and composite materials, et al. The MA and SPS processes can exhibit some significant advantages, such as lower processing temperature, finer grain and more homogeneous microstructure, and excellent mechanical properties of the bulk alloys [7,9].

Our motivation is to develop a HEA material with high strength, closely related to the well-known refractory WTaMoNb HEA [10]. We design $(W_{35}Ta_{35}Mo_{15}Nb_{15})_{95}Ni_5$ RHEA with high theoretical density of 15.12 g/cm^3 via regulating constituent molar ratio and addition of fcc-Ni to accelerate the alloying process. Given the high melting temperature of refractory elements and the traits of fabrication methods, we will adopt powder metallurgical processes to fabricate bulk RHEA.

In this study, a novel $(W_{35}Ta_{35}Mo_{15}Nb_{15})_{95}Ni_5$ RHEA is fabricated by mechanical alloying and spark plasma sintering process. The alloying behavior, phase compositions were analyzed, and the mechanical property was also studied.

2. Materials and Methods

2.1. Synthesis of RHEA

The schematically preparation process was presented in Figure 1. The first process is the synthesis of $(W_{35}Ta_{35}Mo_{15}Nb_{15})_{95}Ni_5$ refractory high entropy alloy (RHEA) powders by mechanical alloying (MA). The RHEA powders were prepared by high-energy ball milling (YXQM-4L, MITR Co., Ltd., Changsha, China) of elemental W, Ta, Mo, Nb and Ni powders, and all the powders were purchased from Beijing Xingrongyuan Technology Co., LTD, China with purity of more than 99.9% and diameter of less than 5 μm. The W, Ta, Mo, Nb and Ni powders with atomic ratio of 33.25:33.25:14.25:14.25:5 were placed in a stainless-steel tank filled with tungsten carbide balls. The MA process was carried out in a high-purity argon gas atmosphere to prevent oxidation with a high-energy planetary ball mill at 300 rpm, and a ball to powder weight ratio of 10:1. To analyze the MA behavior of the HEA powders, powders were sampled after 0.25, 2, 4, 10 and 12 h of milling. All of these procedures were carried out in a glove box filled with high-purity argon gas to prevent oxidation of the HEA powders. In order to ensure the metallic powders fully alloying, the powders ball milled for 12 h were selected to sinter via spark plasma sintering (SPS, HP D25, FCT Systeme GmbH, Rauenstein, Germany) at pressure of 30 MPa under vacuum. The metallic powders were placed in the carbon mold, and then the upper and lower carbon die punches are put into the carbon mold. The SPS process are carried out applying a specific sintering power resource and pressing pressure to the sintered powder. The powders were heated to 420 °C within 2 min, and from 420 °C to final temperature of 1700 °C at a heating rate of 100 °C min^{-1}, followed by 10 min of holding time at the sintering temperature.

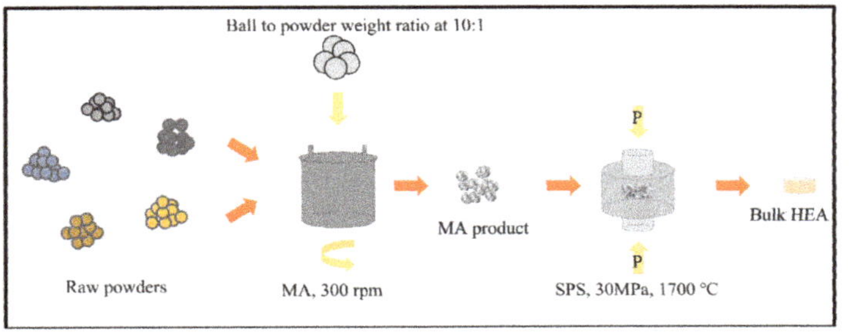

Figure 1. Schematic demonstration of the processing route adopted for the fabrication of $(W_{35}Ta_{35}Mo_{15}Nb_{15})_{95}Ni_5$ high entropy alloy (HEA).

2.2. Characterization

The particle size distribution of the HEA powder was determined by a laser particle size analyzer (MALVERN, MICRO-PLUS, UK) using water as dispersant. The crystal structure of the alloyed powders and sintered bulk specimen were identified by X-ray diffraction (XRD, Rigaku, D/max 2550VB, Japan, scan rate of 8 deg min^{-1}, scan-step size of 0.02 deg, scan angle range of 20–90 deg) using CuK_α radiation at 40 KV and 450 mA. The microstructure and alloying component distribution of the mechanically alloyed HEA powder were analyzed by transmission electron microscopy (TEM, JEM-2100F, JEOL, Tokyo, Japan). The microstructure of the $(W_{35}Ta_{35}Mo_{15}Nb_{15})_{95}Ni_5$ HEA was characterized by field emission scanning electron microscopy (SEM, NOVATM NanoSEM230, FEI, Eindhoven, the Neitherland) equipped with the energy dispersive (EDS, GENESIS60S). The secondary electron (SE) imaging mode was used to observe the morphology of the samples. The back-scattered electron (BSE) imaging mode was used to distinguish the different phases of the microstructure. The alloying element distribution was measured by electron probe microanalysis (EPMA, JXA-8230, JEOR, Tokyo, Japan) with back-scattered mode. Metallographic specimens were prepared by grinding, polishing and chemical etching in an aqueous solution (m_{KOH}:$m_{k3FeCN6}$:m_{H2O} = 1:1:18). The density of the $(W_{35}Ta_{35}Mo_{15}Nb_{15})_{95}Ni_5$ HEA was determined using Archimedes' method with distilled water. Compressive stress–strain testing was carried out in universal testing machine (3369, INSTRON, Bosto, MA, USA) with cylindrical specimens (Φ 6 mm $\times$ 8 mm) at an engineering strain rate of 2 mm min^{-1} at room temperature. The representative data was obtained by averaging three values of the test results.

3. Results and Discussion

3.1. Alloying Process

Figure 2 depicts the XRD patterns of $(W_{35}Ta_{35}Mo_{15}Nb_{15})_{95}Ni_5$ HEA powders after different milling time. The mixed powders after 0.25 h of milling contains diffraction patterns of all constituent elements, while the least atomic weight leaded to that the characteristic peak of intensities of Ni were relatively lower than those of the other elements, because the peak intensities of each element are related to the atomic number of the constituents [11]. The intensities of the diffraction peaks declined sharply and the characteristic peak of Ni disappeared after 2 h of milling, as shown in Figure 2a, which recommends sufficient dissolution of Ni. With the milling time prolonged to 4 h, the characteristic peaks of Mo and Nb clearly merged into the peaks of Ta and W. After 10 h of milling, the constituent elements peaks are no longer observed, and it is obvious that four new diffraction peaks appearing at $2\theta \approx 39.94°$, $57.78°$, $72.56°$ and $86.19°$ are quite well with (110), (200), (211) and (220) planes of a simple solid solution structure. The declined and broadened diffraction peaks are largely ascribed to the refinement of the grain size and the increment of the lattice strain. All diffraction peaks are consistent with a single bcc phase. The minimum alloying time for $(W_{35}Ta_{35}Mo_{15}Nb_{15})_{95}Ni_5$ RHEA powders is shorter than that of some common RHEAs [6,7] powders may due to the addition of Ni.

In fact, the solid solution sequence of the alloying elements is closely related to their melting points [12]. Generally, elements with lower melting points would have higher alloying rate than that with higher melting points. From what has been discussed above, we can draw a conclusion that the mechanical alloying procedure is that the decrease in the peak intensity, broadening and merging of the diffraction peaks and the subsequent disappearance, which is due to the factors: crystal size refinement, lattice strain increasement and reduced crystallinity [13].

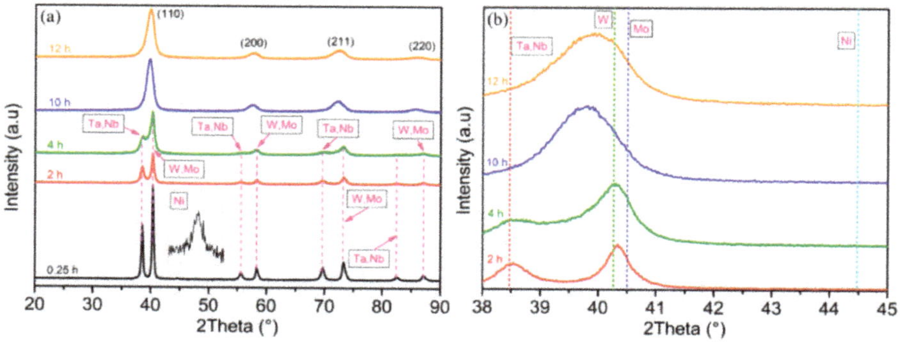

Figure 2. (**a**) XRD patterns of $(W_{35}Ta_{35}Mo_{15}Nb_{15})_{95}Ni_5$ HEA powders after different milling time, and (**b**) magnified XRD patterns of the 2, 4, 10 and 12 h-milled powders at 2θ range of 38–45°.

The breadth of the diffraction peak is a combination of both instrument and sample dependent effects. To eliminate instrumental contributions, it is necessary to obtain a diffraction pattern from the line broadening of a standard material such as silicon powders that are annealed at 1100 °C for 30 min to determine the instrumental broadening. The instrumental-corrected broadening β_D corresponding to the diffraction peak of $(W_{35}Ta_{35}Mo_{15}Nb_{15})_{95}Ni_5$ was evaluated using the relation as follows [14]:

$$\beta_D{}^2 = \beta^2{}_{measured} - \beta^2{}_{instrumental} \tag{1}$$

The grain size was determined according to the Scherrer equation:

$$D = \frac{k\lambda}{\beta_D \cos\theta} \tag{2}$$

where D is the grain size in nanometers, λ is the wavelength of the radiation (0.154056 nm for CuK_α radiation). k is a constant equal to 0.94, β_D is the peak width at half-maximum intensity and θ is the peak position. The lattice strain of HEAs were calculated according to the Williamson–Hall equation [15]:

$$\beta \cos\theta = \frac{k\lambda}{D} + \varepsilon \sin\theta \tag{3}$$

where β is assumed to be β_D, ε is the lattice strain.

The dislocation density ρ was calculated using the following equation [16]:

$$\rho = \frac{2\sqrt{3}\varepsilon}{Db} \tag{4}$$

where ε, D and b are lattice strain, grain size and Burgers vector, respectively. b is equal to $\frac{\sqrt{3}}{2}a$ (a is the lattice parameter determined in the means of the Nelson-Riley approach with the weighting function $1/2(\cot^2\theta + \cos\theta\cot\theta)$ [17]) for a bcc structure, ε and D are calculated from the XRD patterns based on Equation (2).

Figure 3a shows grain size and lattice strain under different milling duration. By and large, with increasing milling time, the grain size become lower, while the lattice strain become larger. As shown, the grain size refined with a large margin to 18.1 nm after 10 h milling time and then slightly decreases to 16.3 nm after 12 h MA, along with the lattice strain increases to 5.87% from 4.15%. Finally, the grain sizes and lattice strain of the powders reaches a dynamic balance after milling for 10 h. The dislocation density gradually become larger with prolonged milling time according to the Equation (4). The morphology and particle size distribution of the HEA powders obtained after 12 h ball milling treatment is shown in Figure 3b. The mechanically alloyed powders with mean size of 5.29 μm are made up of a large amount of refined powders and relatively much larger powders. The approximate spherical morphologies of the HEA powders demonstrate great fluidity, promote the relative high density of as-sintered bulk materials. We choose the powders with milling time of 12 h to sinter, which has been completely alloyed.

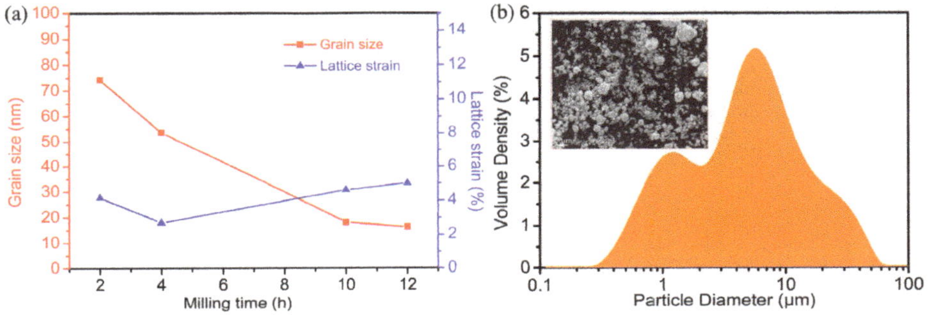

Figure 3. (**a**) Grain size and lattice strain for the powders at different milling time, (**b**) the morphology and the powder particle size distribution curve of the 12 h-milled powders.

Figure 4 is the high-resolution TEM (HRTEM) image of HEA powders after 12 h milling, and fast Fourier transformation images from the region is shown in the inset. The interplanar distance is 0.23 nm determined by the HRTEM image. Scanning TEM-energy dispersive X-ray analysis (STEM-EDX) was also conducted to qualitatively verify the elemental distribution in HEA powder. Figure 5 clearly illustrates that the element of the $(W_{35}Ta_{35}Mo_{15}Nb_{15})_{95}Ni_5$ RHEA are uniformly dispersed.

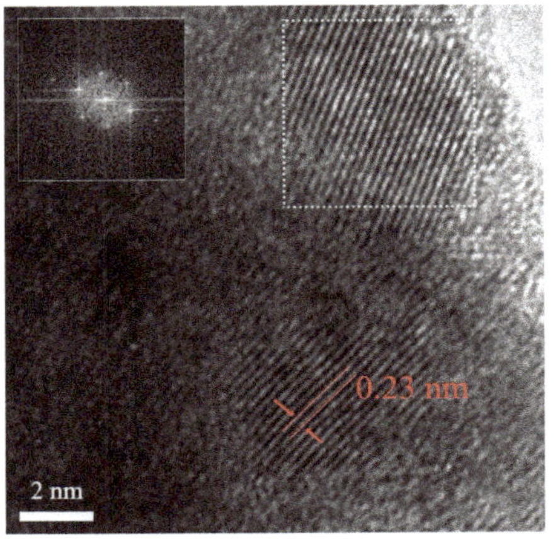

Figure 4. HRTEM image of the HEA powder after 12 h milling and fast Fourier transformation image.

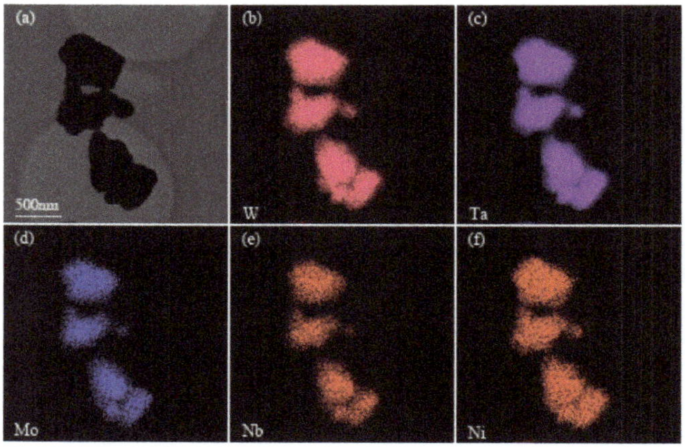

Figure 5. TEM image and corresponding EDX map for each component of HEA powder after 12 h milling, (**a**)TEM image, (**b–f**) EDS mappings of W, Ta, Mo, Nb and Ni, respectively.

3.2. *Phase and Microstructure of the Bulk RHEA*

The crystal structure and phase analysis were investigated by XRD. Figure 6 shows the XRD pattern of the $(W_{35}Ta_{35}Mo_{15}Nb_{15})_{95}Ni_5$ bulk alloy consolidated by SPS. The pattern of the powders milled for 12 h is also presented for comparison. The theoretical crystal lattice parameter a_{mix} of the bcc solid solution in non-equiatomic $(W_{35}Ta_{35}Mo_{15}Nb_{15})_{95}Ni_5$ RHEA is calculated to be a_{mix} = 0.324 nm using the rule of mixtures, while the lattice parameter of as-sintered bulk RHEA determined by the XRD is 0.320 nm based on Nelson-Riley approach [17]. It can be concluded that the calculated value for the $(W_{35}Ta_{35}Mo_{15}Nb_{15})_{95}Ni_5$ RHEA is very close to the experimental result. Compared with the XRD pattern of 12 h-milled powder, peaks shift can be clearly observed for the as-sintered bulk. The peaks position transfer toward lower Bragg angle (2θ), indicating that the lattice parameter of as-sintered bulk RHEA is larger than that of 12 h-milled RHEA powder (a = 0.319 nm). The XRD patterns indicate that the bulk specimens exhibit a single bcc structure. Besides, the minor diffraction peaks in the XRD pattern closely matched with cubic $Nb_{5.7}Ni_4Ta_{2.3}O_2$ (PDF#82-0947). A few diffraction peaks intensity is inconsistent with that of the PDF card may because the intensities of the XRD diffraction peaks decreases due to the appeared extremely refined oxide inclusion. As shown in Figure 6, the predominant crystal structure for both $(W_{35}Ta_{35}Mo_{15}Nb_{15})_{95}Ni_5$ bulk sample and powders is a single-phase solid solution with bcc structure, which demonstrates there was no phase decomposition during the SPS process.

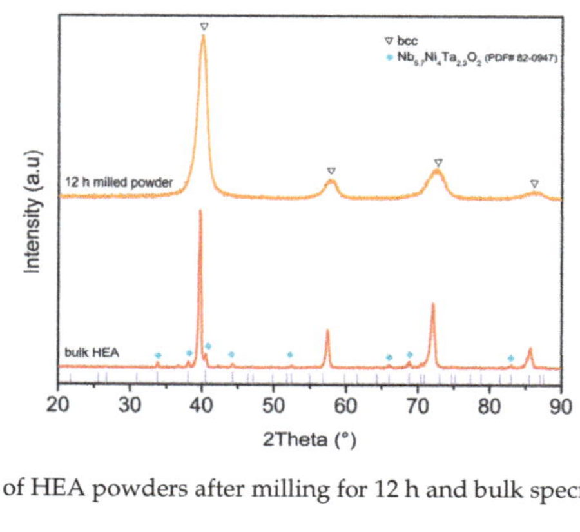

Figure 6. XRD patterns of HEA powders after milling for 12 h and bulk specimen sintered at 1700 °C.

Generally, the formation of metallic materials decides their physical and chemical properties. High entropy alloys are deemed as a kind of novel metallic materials, thus the phase structure and stability prediction have great significance for the design and application of these types of material. Since the concept of HEA was introduced by Yeh et al. [18], many literatures have reported physical parameters to predict the phase structure and stability of HEAs. Especially, the parameters δ (the atomic size difference) and Ω (a thermodynamic parameter) that were proposed [19] can predict the structure and stability of HEA. The parameter δ is defined as:

$$\delta = \sqrt{\sum_{i=1}^{n} c_i\left(1 - \frac{ri}{\bar{r}}\right)^2} \tag{5}$$

where c_i is the atomic fraction of the ith element, r_i is the atomic radius of the ith element and $\bar{r}$ is the average atomic radius, which is calculated by $\bar{r} = \sum_{i=1}^{n} c_i r_i$.

The other parameter (Ω) is defined as:

$$\Omega = \frac{T_m \Delta S_{mix}}{|\Delta H_{mix}|} \tag{6}$$

where T_m is melting point calculated as weighted average of the constituting elements, ΔS_{mix} is the entropy of mixing for HEAs, which is calculated by $\Delta S_{mix} = -R \sum_{i=1}^{n} c_i \ln c_i$ and ΔH_{mix} is the enthalpy of mixing of HEAs, which is calculated by $\Delta H_{mix} = \sum_{i=1, i\neq j}^{n} 4\Delta H_{mix}{}^{AB} c_i c_j$. According to Yang [19], as $\delta \leq 6.6\%$ and $\Omega \geq 1.1$, a stable, solid solution phase is constructed in a HEA system.

The literature has proposed another important parameter [20], the valence electron concentration (VEC). VEC is calculated by:

$$VEC = \sum_{i=1}^{n} c_i (VEC)_i \tag{7}$$

where $(VEC)_i$ is the VEC for the ith alloying element. Guo et al. [20] criticized that a low VEC (<6.87) is partial to sole bcc structure's solid solution phase, whereas a high VEC (≥8) favors sole solid solution phase with fcc structure. When $6.87 \leq VEC < 8$, the mixture of bcc and fcc solid solution phase is displayed.

According to the Equations (5)–(7), the calculated values of δ, Ω and VEC for the $(W_{35}Ta_{35}Mo_{15}Nb_{15})_{95}Ni_5$ HEA are 3.27%, 4.16 and 5.73 respectively. Hence, the formation of the current HEA is supposed to be a formation of single-phase solid solution with bcc structure. Our experimental results coincided exactly with the prediction.

Figure 7a shows the SEM image of the as-sintered $(W_{35}Ta_{35}Mo_{15}Nb_{15})_{95}Ni_5$ HEA. Likewise, the back-scatter electron imaging mode was utilized to distinguish the different phases of the bulk HEA as shown in Figure 7b. It suggests that the $(W_{35}Ta_{35}Mo_{15}Nb_{15})_{95}Ni_5$ HEA is composed of a light gray matrix and dark gray inclusion. We adopt SEM-EDS to determine chemical components in region A and region B. The EDS results are presented in Figure 7c,d. The EDS analysis result presents the region B were enriched in Ta, Nb and O, and had a much higher Ni content than region A, while the W, Mo elements in region B are deficient.

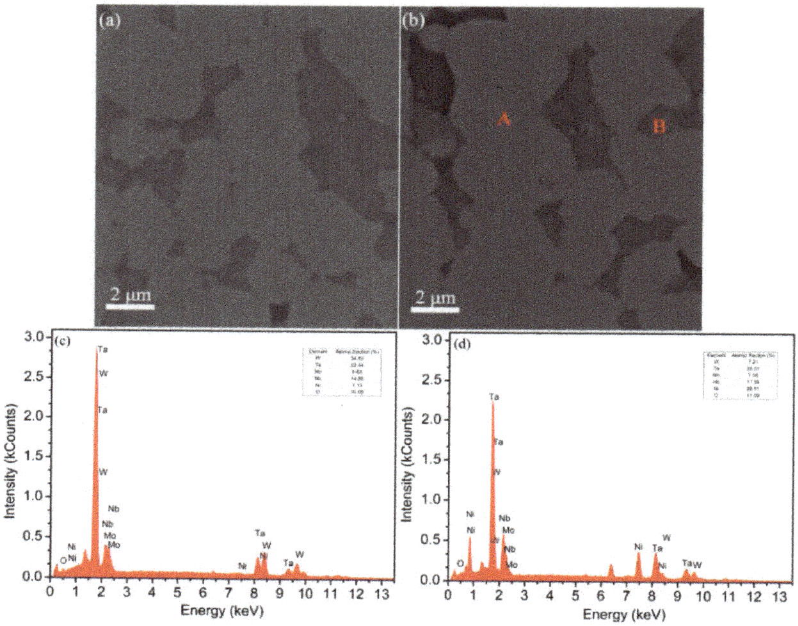

Figure 7. SEM image of the bulk $(W_{35}Ta_{35}Mo_{15}Nb_{15})_{95}Ni_5$ RHEA (**a**,**b**) with secondary electron and back-scattering electron mode, respectively, and EDS analysis of (**c**) region A and (**d**) region B marked in (**b**).

In order to further analyze the composition of the second phase in the bulk alloy, the back-scatter electron imaging mode and corresponding EDS-mapping were used to measure the distribution of the alloying elements in inclusion and matrix. The EDS-mapping analysis results are shown in Figure 8. The distribution of Ta, Nb elements was homogeneous in the HEA, whereas, the distribution of W, Mo, Ni elements were inhomogeneous. The light matrix is enriched in W, Mo and the distribution of W, Mo elements are homogeneous, but in the dark gray region are deficient. The dark gray region is enriched in Ni element, but in the light grey matrix were deficient. These results are consistent with the XRD analysis that the dark gray oxide inclusion should be $Nb_{5.7}Ni_4Ta_{2.3}O_2$ phase. The microstructure and element distribution of the alloy was further confirmed by EPMA. Figure 9 presents the EPMA mapping of alloying elements of the $(W_{35}Ta_{35}Mo_{15}Nb_{15})_{95}Ni_5$ HEA. As can be seen, alloying components Ni, Nb and Ta content is rich, while W and Mo are extremely minor in the region demonstrated by yellow circle. Thus, the EPMA analysis agrees with the XRD and the SEM results.

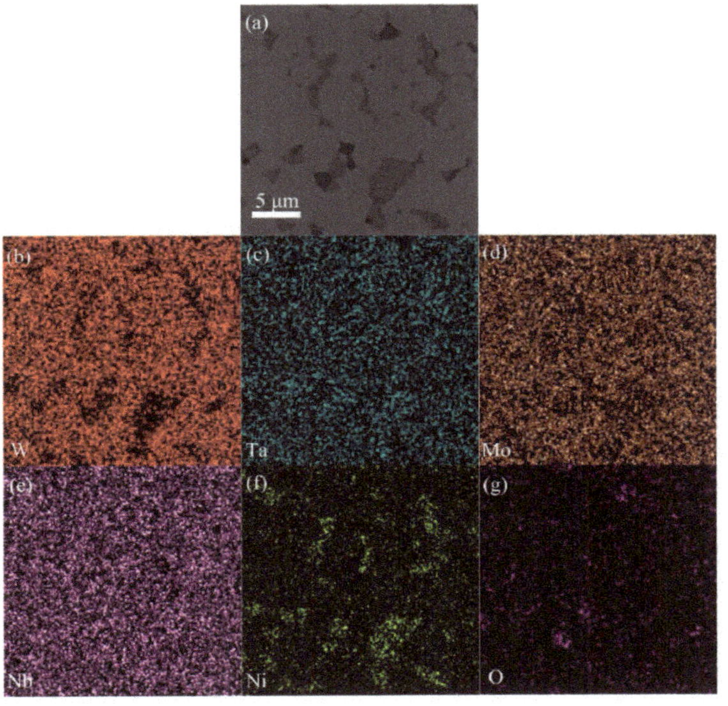

Figure 8. SEM-BSE image and corresponding EDS map for each component of $(W_{35}Ta_{35}Mo_{15}Nb_{15})_{95}Ni_5$ HEA bulk specimen, (**a**) SEM image, (**b**–**g**) EDS mappings of W, Ta, Mo, Nb, Ni and O respectively.

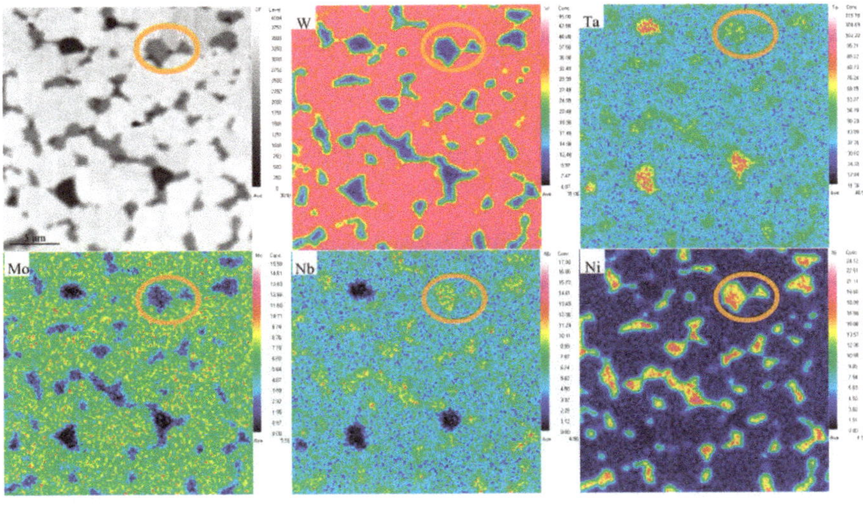

Figure 9. The EPMA-BSE image and the related mapping of alloying elements of the $(W_{35}Ta_{35}Mo_{15}Nb_{15})_{95}Ni_5$ HEA.

Figure 10a shows the microstructure of the $(W_{35}Ta_{35}Mo_{15}Nb_{15})_{95}Ni_5$ HEA bulk after chemical etching. The bulk materials via powder metallurgical process have a lot of pores as marked by red arrows. From the image, we can see that a large amount of inclusion marked by yellow arrows, are uniformly dispersed in the grain boundaries of the matrix combing with Figure 7b. We can clearly observe the grain marked by red dotted line, and the grain size of the bulk HEA was also measured from Figure 10a by nano-measurer software. Figure 10b depicts the statistical distribution of grain size and its Gaussian fitting. From Figure 10b, we can see that the grain size mainly ranges from 2.25 to 4.25 μm, and the average grain size is around 3.25 μm.

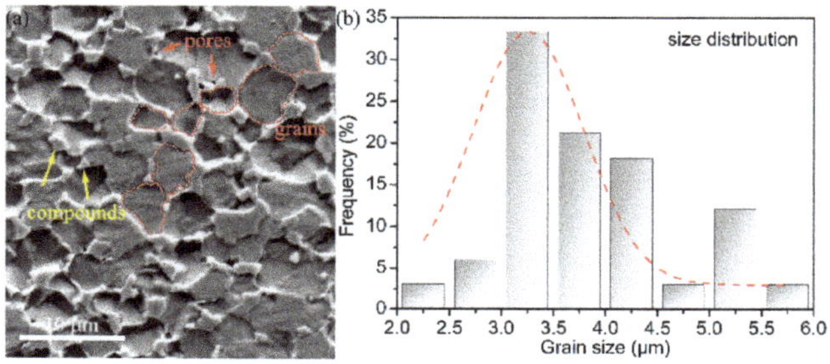

Figure 10. (**a**) SEM image of the bulk $(W_{35}Ta_{35}Mo_{15}Nb_{15})_{95}Ni_5$ RHEA after etching, (**b**) grain size distribution of HEA matrix.

3.3. *Mechanical Properties*

Compression test was performed at room temperature. Figure 11 shows the engineering compression stress–strain curve for the $(W_{35}Ta_{35}Mo_{15}Nb_{15})_{95}Ni_5$ RHEA. The compressive mechanical property of the bulk $(W_{35}Ta_{35}Mo_{15}Nb_{15})_{95}Ni_5$ RHEA and typical refractory HEAs reported were summarized in Table 1. In contrast with other typical RHEAs, the $(W_{35}Ta_{35}Mo_{15}Nb_{15})_{95}Ni_5$ RHEA possesses the advantages of both high strength and high fracture strain. It is necessary to take the microstructure and crystal structure of each phase into account to analyze the factors of the outstanding mechanical properties for the $(W_{35}Ta_{35}Mo_{15}Nb_{15})_{95}Ni_5$ RHEA. Firstly, solid solution strengthening is an intrinsic concept of HEAs on account of a severe lattice distortion [2]. Furthermore, unlike the typical as-casted RHEAs that have larger coarse-grains (grain size: 50–100 μm) [21], the as-sintered $(W_{35}Ta_{35}Mo_{15}Nb_{15})_{95}Ni_5$ RHEA displays a much finer and more homogeneous microstructure (grain size was approximately 3.25 μm), implying that grain boundary strengthening play a dominant part in the strength improvement. Grain boundary strengthening mechanism is described by the Hall–Petch formula ($\Delta\sigma_{gb} = k_y/\sqrt{d}$, where k_y is the Hall–Petch coefficient and d is the average grain size). In addition, it should be noted that the $(W_{35}Ta_{35}Mo_{15}Nb_{15})_{95}Ni_5$ RHEA in this study has marginal volume fraction of oxide inclusion phases $Nb_{5.7}Ni_4Ta_{2.3}O_2$ which contribute to the precipitation strengthening. Generally, precipitation strengthening mechanism complies with Orowan dislocation bypass mechanism or dislocation shearing mechanism. Given the two mechanisms traits, the high shear modulus and hardness of the oxide inclusion, Orowan dislocation bypass mechanism is more appropriate to assess the contribution of precipitate-reinforce mechanism. Finally, it is inevitable that interstitial elements such as O, N, C, etc, are introduced during the MA and SPS process, which contributes to interstitial solid solution strengthening [22]. Figure 11b,c depict compressive fracture morphologies of the $(W_{35}Ta_{35}Mo_{15}Nb_{15})_{95}Ni_5$ RHEA at room temperature. We can deduce that the compressive fracture mode is cleavage fracture from the fracture morphologies present a mixture of river-like pattern and steps.

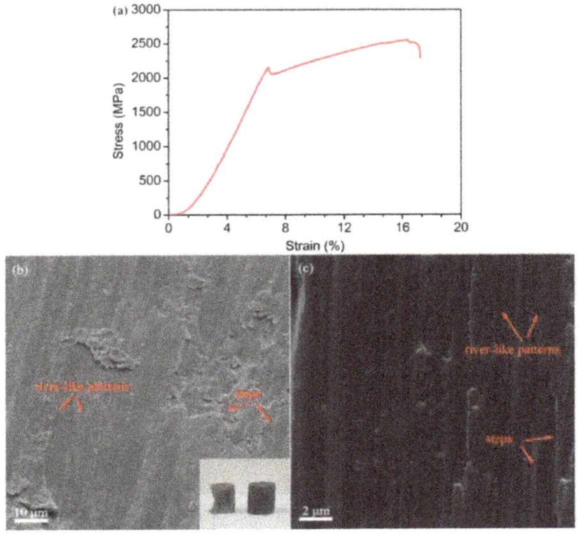

Figure 11. (**a**) Engineering stress–strain curve of the bulk $(W_{35}Ta_{35}Mo_{15}Nb_{15})_{95}Ni_5$ RHEA at room temperature, (**b**) fracture surface morphology and inserted compression sample image and (c) magnified fracture morphology.

Table 1. Compressive mechanical properties of typical RHEAs at room temperature.

Alloy	Process	Phase	$\sigma_{0.2}$(MPa)	σ_{max}(MPa)	ε_f(%)	Refs.
$(W_{35}Ta_{35}Mo_{15}Nb_{15})_{95}Ni_5$	MA+SPS (1700 °C)	bcc	~2128	~2562	8.16	This study
WTaMoNb	Arc melting	bcc	1058	1211	1.5	[10]
WTaMoNbV	Arc melting	bcc	1246	1270	1.7	[10]
WTaMoRe	Arc melting	bcc	-	1451	5.69	[23]
WTaMoNb	SPS	bcc	1217	1499	3.8	[24]
$WTaMoNbSi_{0.25}$	SPS	bcc	1826	2548	10.5	[24]
$WTaMoNbSi_{0.5}$	SPS	bcc	1883	2454	5.8	[24]

To sum up, the as-sintered $(W_{35}Ta_{35}Mo_{15}Nb_{15})_{95}Ni_5$ RHEA may be regarded as a potential candidate for high temperature structural material with high density and outstanding compressive strength owing to a combination of the four strengthening mechanisms.

4. Conclusions

A novel high strength $(W_{35}Ta_{35}Mo_{15}Nb_{15})_{95}Ni_5$ RHEA with high density was successfully fabricated by MA and SPS process in this study. The 12 h-milled powders with an average crystalline size of about 16.3 nm exhibit a single bcc solution phase. After the SPS process at 1700 °C, the bulk RHEA has a relative density of exceeding 96% and the average grain size is 3.25 µm. Meanwhile, minor oxide inclusion $(Nb_{5.7}Ni_4Ta_{2.3}O_2)$ are detected in the bulk sample. The compressive yield stress, maximum stress and fracture strain of the $(W_{35}Ta_{35}Mo_{15}Nb_{15})_{95}Ni_5$ RHEA are 2128 MPa, 2562 MPa and 8.16%, respectively, which are superior to some typical RHEAs. The extraordinary high mechanical properties are attributed to the following factors: solid solution strengthening due to a severe lattice distortion, which is an intrinsic concept of HEAs, grain boundary strengthening, precipitation strengthening mechanism by second phases.

Author Contributions: Investigation, B.D. and Y.Y.; methodology, D.W.; project administration, X.L.; writing—original draft, Y.Y.; writing—review & editing, Z.W. All authors have read and agreed to the published version of the manuscript.

Funding: This research was supported by the Natural Science Foundation of Hunan Province, China (No. 2018JJ3677).

Conflicts of Interest: The authors declare that they have no known competing financial interests or personal relationships that could have appeared to influence the work reported in this paper.

References

1. Leek, H.; Chas, I.; Ryuh, J.; Hong, S.H. Effect of two-stage sintering process on microstructure and mechanical properties of ODS tungsten heavy alloy. *Mater. Sci. Eng. A* **2007**, *458*, 323–329.
2. Zhang, Y.; Yang, X.; Liaw, P.K. Alloy Design and Properties Optimization of High-Entropy Alloys. *JOM* **2012**, *64*, 830–838. [CrossRef]
3. Senkovo, N.; Miracled, B.; Chaputk, J. Development and exploration of refractory high entropy alloys—A review. *J. Mater. Res.* **2018**, *33*, 3092–3128. [CrossRef]
4. Li, Z.; Pradeep, K.G.; Deng, Y.; Raabe, D.; Tasan, C.C. Metastable high-entropy dual phase alloys overcome the strength-ductility trade-off. *Nature* **2016**, *534*, 227–230. [CrossRef] [PubMed]
5. Li, B.S.; Wang, Y.P.; Ren, M.X.; Yang, C.; Fu, H.Z. Effects of Mn, Ti and V on the microstructure and properties of AlCrFeCoNiCu high entropy alloy. *Mater. Sci. Eng. A* **2008**, *498*, 482–486. [CrossRef]
6. Wang, G.; Liu, Q.; Yang, J.; Li, X.; Sui, X.; Gu, Y.; Liu, Y. Synthesis and thermal stability of a nanocrystalline MoNbTaTiV refractory high-entropy alloy via mechanical alloying. *Int. J. Refract. Met. Hard Mater.* **2019**, *84*, 104988. [CrossRef]
7. Long, Y.; Liang, X.; Su, K.; Peng, H.; Li, X. A fine-grained NbMoTaWVCr refractory high-entropy alloy with ultra-high strength: Microstructural evolution and mechanical properties. *J. Alloy Compd.* **2019**, *780*, 607–617. [CrossRef]
8. Liu, Y.; Lu, Y.; Zhang, Y.; Miao, W. Study of nickel's effect on the synthesis of TiC during mechanical alloying. *Powder Metall. Technol.* **2007**, *25*, 96–98. (In Chinese)
9. Kang, B.; Lee, J.; Ryu, H.J.; Hong, S.H. Ultra-high strength WNbMoTaV high-entropy alloys with fine grain structure fabricated by powder metallurgical process. *Mater. Sci. Eng. A* **2018**, *712*, 616–624. [CrossRef]
10. Senkov, O.N.; Wilks, G.B.; Scott, J.M.; Miracle, D.B. Mechanical properties of Nb25Mo25Ta25W25 and V20Nb20Mo20Ta20W20 refractory high entropy alloys. *Intermetallics* **2011**, *19*, 698–706. [CrossRef]
11. Cullity, B.D.; Stock, S.R. *Elements of X-ray Diffraction*, 3rd ed.; Prentice Hall: Upper Saddle River, NJ, USA, 2001.
12. Chen, Y.L.; Hu, Y.H.; Hsieh, C.A.; Yeh, J.W.; Chen, S.K. Competition between elements during mechanical alloying in an octonary multi-principal-element alloy system. *J. Alloy Compd.* **2009**, *481*, 768–775. [CrossRef]
13. Ji, W.; Wang, W.; Wang, H.; Zhang, J.; Wang, Y.; Zhang, F.; Fu, Z. Alloying behavior and novel properties of CoCrFeNiMn high-entropy alloy fabricated by mechanical alloying and spark plasma sintering. *Intermetallics* **2015**, *56*, 24–27. [CrossRef]
14. Zak, A.K.; Majid, W.A.; Abrishami, M.E.; Yousefi, R. X-ray analysis of ZnO nanoparticles by WilliamsoneHall and sizeestrain plot methods. *Solid State Sci.* **2011**, *13*, 251–256.
15. Williamson, G.K.; Hall, G.K. X-ray line broadening from filed aluminium and wolfram. *Acta Metall.* **1953**, *1*, 22–31. [CrossRef]
16. Williamson, G.K.; Smallman, R.E. Dislocation densities in some annealed and cold-worked metals from measurements on the X-ray debye-scherrer spectrum. *Philos. Mag.* **1956**, *1*, 34–46. [CrossRef]
17. Schliephake, D.; Medvede, A.E.; Imran, M.K.; Obert, S.; Fabijanic, D.; Heilmaier, M.; Molotinkov, A.; Wu, X.H. Precipitation behaviour and mechanical properties of a novel Al0.5MoTaTi complex concentrated alloy. *Scr. Mater.* **2019**, *173*, 16–20. [CrossRef]
18. Yeh, J.W.; Chen, S.K.; Lin, S.J.; Gan, J.Y.; Chin, T.S.; Shun, T.T.; Tsau, C.H.; Chan, G.S.Y. Nanostructured high-entropy alloys with multiple principle elements: Novel alloy design concepts and outcomes. *Adv. Eng. Mater.* **2004**, *6*, 299–303. [CrossRef]
19. Yang, X.; Zhang, Y. Prediction of high-entropy stabilized solid-solution in multi-component alloys. *Mater. Chem. Phys.* **2012**, *132*, 233–238. [CrossRef]

20. Guo, S.; Ng, C.; Lu, J.; Liu, C.T. Effect of valence electron concentration on stability of fcc or bcc phase in high entropy alloys. *J. Appl. Phys.* **2011**, *109*, 645–647. [CrossRef]
21. Lv, S.; Zu, Y.; Chen, G.; Fu, X.; Zhou, W. An ultra-high strength CrMoNbWTi-C high entropy alloy co-strengthened by dispersed refractory IM and UHTC phases. *J. Alloy Compd.* **2018**, *788*, 1256–1264. [CrossRef]
22. Long, Y.; Zhang, H.; Wang, T.; Huang, X.; Li, Y.; Wu, J.; Chen, H. High-strength Ti–6Al–4V with ultrafine-grained structure fabricated by high energy ball milling and spark plasma sintering. *Mater. Sci. Eng. A* **2013**, *585*, 408–414. [CrossRef]
23. Wei, Q.; Shen, Q.; Zhang, J.; Chen, B.; Luo, G.; Zhang, L. Microstructure and mechanical property of a novel ReMoTaW high-entropy alloy with high density. *Int. J. Refract. Met. Hard Mater.* **2018**, *77*, 8–11. [CrossRef]
24. Guo, Z.; Zhang, A.; Han, J.; Meng, J. Effect of Si additions on microstructure and mechanical properties of refractory NbTaWMo high-entropy alloys. *J. Mater. Sci.* **2019**, *54*, 5844–5851. [CrossRef]

Publisher's Note: MDPI stays neutral with regard to jurisdictional claims in published maps and institutional affiliations.

metals

MDPI

Article

Effects of the Addition of Fe, Co on the Azo Dye Degradation Ability of Mn-Al Mechanically Alloyed Powders

Wael Ben Mbarek [1], Joan Saurina [1], Lluisa Escoda [1], Eloi Pineda [2], Mohamed Khitouni [3] and Joan-Josep Suñol [1,*]

1 Department of Physics, Campus Montilivi, s/n University of Girona, 17003 Girona, Spain; u1930157@correu.campus.udg.edu (W.B.M.); joan.saurina@udg.edu (J.S.); lluisa.escoda@udg.edu (L.E.)

2 Department of Physics and nuclear Engineering, Campus Diagonal Besòs, Polytechnic University of Catalonia, 08019 Barcelona, Spain; eloi.pineda@upc.edu

3 Inorganic Chemistry Laboratory, UR-11-ES-73, University of Sfax, Sfax BP 1171, Tunisia; mohamed.khitouni@fss.rnu.tn

* Correspondence: joanjosep.sunyol@udg.edu; Tel.: +34-972-419-757

Received: 15 October 2020; Accepted: 22 November 2020; Published: 25 November 2020

Abstract: Azo compounds are used in the textile and leather industry. A significant step during the azo dyes treatment of water is the degradation by breaking the N=N bonds. This break produces the decolorization of water. In this research work, 10% atomic of Fe or Co was added to produce ternary Mn-Al-rich, nanostructured, mechanically alloyed powders in order to improve the decolorization of Reactive Black 5 solutions and to check Fe and Co addition's influence. The microstructure was followed by X-ray diffraction, the morphology and composition by electronic microscopy and energy-dispersive X-ray spectroscopy (EDS) microanalysis. The dye degradation was monitored with ultraviolet/visible absorption spectrophotometry. After degradation, the remaining organic compound was checked by high-performance liquid chromatography (HPLC) and the functional groups of the powdered alloys by infrared spectroscopy. Fe addition to Mn-Al displayed faster kinetics and a higher efficiency than the Co addition. The Mn-Al-Fe solution (0.25 g/100 mL) was fully decolorized in 5 min. On the other side, Mn-Al-Co powders were able to successfully decolorize the dyed solution in 10 min under the same conditions. Thus, nanocrystalline Fe-doped Mn-Al alloys are good candidates for use in the decolorization process, in comparison with Co-doped and other intermetallic particles.

Keywords: mechanical alloying; reactive black 5; decolorization; UV-visible spectrophotometry; LC-MS analysis

1. Introduction

The industrial water effluent is one important source of pollution in the environment [1]. The dye effluents from textile industries are significant sources of pollution (high toxicity, low biodegradation rate) [2]. Various processes and materials have been proposed in the scientific literature for wastewater treatment. Some examples are: adsorption onto granular activated carbon [3], hollow fiber ultrafiltration [4], coagulation-flocculation with Detarium microcarpum [5], biological oxidation [6], or some photocatalyst methods and materials [7,8] including a magnetic photocatalyst [9]. One of the methods is reduction with zero valent metals or alloys [10]. They have been applied for removing azo dyes [7] as well as metals and chlorinated organics [11–14].

The metallic particles (MPs) have been applied as elements or as alloys to facilitate the reduction process. The microstructure of the particles can be amorphous or crystalline. Usually, the best results are found in amorphous because the corrosion resistance and chemical interaction are usually improved if the comparison is performed with crystalline alloys with the same composition [15]. Thus, as microstructure affects the functional behavior, in this work, we analyze some nanocrystalline Mn-rich alloys. Nanocrystalline materials have a crystalline structure, and between 10 and 50% of atoms are in the grain boundaries between crystalline grains. In the scientific literature, there are interesting works about the degradation induced by MPs in dyes (by chemical reaction, breaking bonds). One of the main topics is the analysis of the degradation reaction by taking into account the particle's surface activity (with the organic contaminants) of the metallic particles [16,17]. The reaction of degradation of the dye molecules is based on a redox process. On the surface of the metallic alloys, the metallic atoms lose electrons. These electrons are effective to cleave the active bonds (as the azo dye -N=N- bonds) of organic molecules [18–22].

There are some methods that help to develop metallic particles: precipitation (magnetite, iron) [23,24], gas atomization (high-entropy alloys) [25] or mechanical alloying (Mn-Al-based alloys) [19,22]. In the mechanical alloying (or ball milling) process, particles yield severe plastic deformation, producing stress and strains. Thus, the mechanical deformation favors the development of metastable microstructures (amorphous, nanocrystalline) and, subsequently, the mechanical and functional properties are different from crystalline alloys with the same composition [26].

In this work, we analyze the azo dye degradation ability by improving the rate of the degradation reaction of ternary Mn-Al- (Fe, Co) nanocrystalline powders produced by mechanical alloying. A high degradation rate will favor the integration of this material as a preliminary step procedure for the full degradation of azo dye wastewaters.

In previous works, we reported the high efficiency of Manganese-Aluminum (Mn-Al) particles in the degradation of azo-dyes: reactive black 5 (RB5) [19] and orange II [21]. It should be remarked that Mn and Al (as biocompatible elements) are used in environmental and biomedical applications [27].

2. Materials and Methods

The $Mn_{60}Al_{30}Fe_{10}$ and $Mn_{60}Al_{20}Co_{10}$ (at.%) alloys were produced from elemental high-purity (<99.9 at.%) Mn, Al, Fe or Co by mechanical alloying (MA) under an Ar atmosphere for 15 h at 500 rpm. Cycles (10 min milling/5 min off) are applied to prevent the excessive heating, with the inversion of the rotation in each cycle (to prevent powders sticking to milling media).

The morphology, composition, surface area, and microstructure of the powders were explored with scanning electron microscopy (SEM) at 15 kV. Being integrated in the SEM equipment, microanalysis was performed by energy-dispersive X-ray spectroscopy (EDS, Vega©Tescan, Brno, Czech Republic). The reactivity is linked to the specific surface area of the Mn-Al- (Fe, Co) powders. This specific area is determined by the gas multilayer adsorption method in a BET device (Micromeritics ASAP 2010 M, Norcross, GA, USA) under nitrogen, after degassing the powder at 300 °C for 24 h. The microstructure was determined at room temperature by X-ray diffraction (XRD, Siemens/Bruker D500, Billerica, MA, USA) using Cu-K_{α} radiation.

In order to assess the colorant (RB5) degradation reaction, an aqueous solution (dye concentration: 40 mg L^{-1}). The procedure was described and portrayed in previous works [19,20]. The produced supernatants were separated, and the color intensity was measured at the maximum absorption wavelength by ultraviolet-visible absorption spectrophotometry (UV-Vis, Shimadzu 2600, Kyoto, Japan), wavelength scan in the range between 200 and 800 nm).

To check whether these materials can be used in the integrated treatment of wastewater, it is necessary to analyze the metallic powders after degradation as well as the remaining wastewater liquid (the procedure described in references 19 and 20). The powder collected after degradation was analyzed using Fourier transform infrared spectroscopy (FTIR) (Bruker, Billerica, MA, USA) and the concentrations of the metal ions checked by absorption spectrophotometer according to standard methods [19]. The organic compounds from the RB5 degradation were analyzed by high-performance liquid chromatography (HPLC) (Agilent, Santa Clara, CA, USA) coupled with tandem mass spectrometry (LC-MS/MS, Bruker, Billerica, MA, USA) in a Beckman Gold chromatograph fitted with a Proshell 120 Pheny Hexyl column at room temperature.

3. Results

First, we check powders morphology and composition. Figure 1a,b displays the surface morphology of the MA Mn-Al-Fe powder. As exhibited in the figure, many corrugations appear on the surface of the nanostructured powders. The EDS microanalysis finding reported in Figure 1c does not reveal significant contamination by the production process (milling tools) and the MA powder is mainly composed (in the volume) of the precursors. The C is due to the sputtering process used to prepare samples for SEM observation. Oxygen contamination is typical of powdered metals with a high surface/volume ratio due to contact with the air before and after the MA process in the Ar atmosphere. Furthermore, oxides were undetected in the powders. The atomic concentration of the elements (Mn, Al and Fe) is 57:31:12, These values are similar to those of the nominal composition: 60:30:10. It should be remarked that EDS is considered as a semi-quantitative method. The size distribution of the powders (by neglecting big powders associated with conglomerates) is provided in Figure 1d (mean particle size is $\approx$10 μm, the accuracy of the size is 0.5 μm and the standard deviation 3.6 μm). The micrographs of the Mn-Al-Co as-produced powder are illustrated in Figure 2a,b. As expected in ball-milled alloys, smooth surfaces, particle aggregation and corrugations were found. The EDS microanalysis (Figure 2c) proves that the Mn:Al:Co ratio was approximately equal to 55:34:11, similar to the 60:30:10 nominal composition. Likewise, the distribution of the particles size is shown in Figure 2d. For statistical comparison, the particle size distribution was also analyzed. The average particle size found to be slightly lower than those of Mn-Al-Fe particles, at $\approx$9 μm (this slight difference can be influenced by the particle selection manual procedure for SEM analysis), and the standard deviation is 1.9 μm.

The decolorization was checked by analyzing the solutions described in the materials and methods section. Figure 3a presents the image of RB5 solutions before and after the addition of MA powders (25 °C). It is clear that the discoloration of the RB5 aqueous solution RB5 is very successful. The supernatants of the aqueous solution of RB5 treated with the two ternary powders Mn-Al-Fe and Mn-Al-Co were separated, respectively, at 0.5, 1, 2, 3, 4 and 5 min as well as 0, 1, 5, 10, 15 and 20 min. Figure 3b,c shows the evolution of absorbance of different aqueous supernatant portions which were measured by UV-Vis absorption spectrophotometry. For the ternary compound Mn-Al-Fe, the solution is fully discolored in 5 min (Figure 3b), whereas the solution treated with the Mn-Al-Co powder is fully discolored in 20 min (Figure 3c). The peak at 597 nm of the absorbance is associated to the azo dye concentration in the solution [19,20,28]. This peak in the visible region is associated to the "-N=N-" bonds and to the dye molecule concentration in the solution. Thus, the decrease in the intensity (as the reaction time in the solution increases) at 597 nm is an indicator of the degradation and RB5 chromophores' evolution. Likewise, the bands (at 230 and 310 nm) in the ultraviolet region are assigned, respectively, to the benzene and naphthalene rings of the molecule dye [19,20,29,30]. These bands, at λ_{max}, become weaker with degradation time. This effect is linked to: (a) the cleavage of the azo bands (also confirmed by the increase of 246 nm in the absorbance peak), (b) the formation of (-NH_2) groups, (c) the decomposition of RB5 in the solution. This result was also found by Zhang and coworkers [31], who confirmed the reductive degradation by amorphous Fe of azo acid orange II solution and Ben Mbarek et al. [19,20], who highlighted the rapid degradation of azo-dye with mechanically alloyed nanocrystalline Mn-Al alloys, as well as the high-efficiency decolorization of azo dyes by Ca-Al particles.

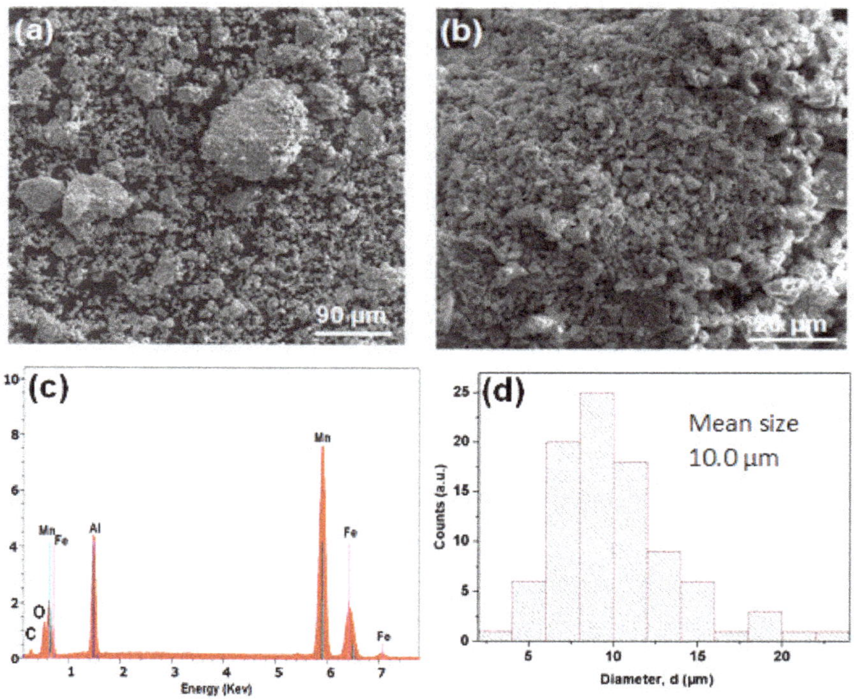

Figure 1. (**a**,**b**) Micrographs of the mechanical alloying (MA) Mn-Al-Fe. (**c**) Energy-dispersive X-ray spectroscopy (EDS) microanalysis and (**d**) distribution of particle sizes of the mechanically alloyed powder.

The mechanism of the redox reaction is viewed to be practically the same as for the case of the binary Mn-Al. However, the rates of RB5 bleaching reaction under the same dosage and temperature conditions using the Mn-Al-Fe and Mn-Al-Co powders appear to be different. The different potentials of the reduction in Fe and Co could account for these discrepancies, as well as the different solubilities of their hydroxides. It is known that the standard oxidation potential Co/Co^{2+} is less negative than the oxidation potential of H^+/H_2, whereas the standard oxidation potential of Fe/Fe^{2+} is more negative (−0.29 and −0.44 V, respectively), which makes the transfer of electrons between Co and H^+ slower than that occurring between the Fe and H^+ [32]. For the improvement in the azo bond-breaking, Co/Fe are active actors favoring the production of hydrogen from water [31,33–35].

Another factor to consider is the different adsorption capacity of both elements. The valence electrons (metal atoms) are shared by all atoms to form a cloud of electrons in a highly shared state. The valence electron configurations are $3d^6 4s^2$ (Fe) and $3d^7 4s^2$ (Co). Magnetic data showed that there was an average of 2.2 and 1.7 holes in the d band for Fe and Co, respectively [36]. The more holes in the bands, the more unpaired electrons there are. Therefore, the interaction between unpaired electrons and adsorbate molecules favors the formation of a localized adsorption bond. Thus, higher adsorption capacity is linked to the unpaired electrons. Simultaneously to the adsorption of the RB5 molecule, a reduction in the azo group might take place on the surface of the metallic powder. Thus, the selected value to follow the kinetics reaction was the maximum absorption wavelength. The compound Mn-Al-Fe exhibits the highest decolorization process reaction rate in the first five minutes, since it has the highest concentration of d-band holes per unit atom [36]. However, the compound Mn-Al-Co displays a lower adsorption rate (Figure 3d).

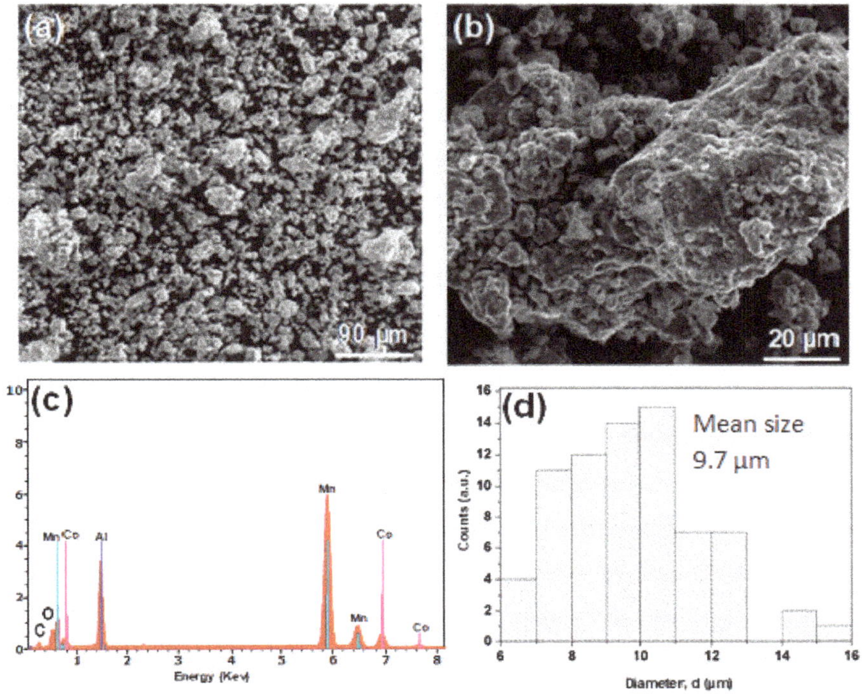

Figure 2. (**a**,**b**) Particle morphologies of the MA Mn-Al-Co powder. (**c**) EDS microanalysis and (**d**) distribution of particle sizes of the ball milled powder.

As shown in Figure 4, a straight line of $\ln(C/C_o)$ versus reaction time was achieve, with coefficients of determination R^2 (0.9322 and 0.9813, values > 0.9). Therefore, the decolorization of R5 solution is consistent with first-order kinetics. The slope of the alloy with Fe addition is 0.9725 s^{-1}, higher than 0.1717 s^{-1} (Mn-Al-Co). The Mn-Al-Co value is also higher than those of Mn-Al (0.2154 s^{-1}). Thus, Co addition does not provoke a significant change in the decolorization efficiency, whereas Fe addition favors the process.

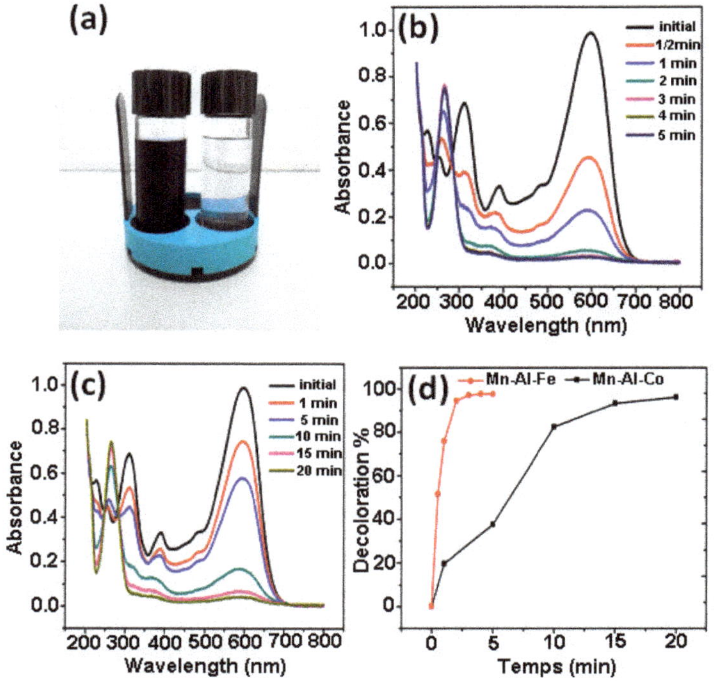

Figure 3. (**a**) Image of black 5 solutions processed by MA powder before and after degradation. (**b**) UV absorption spectra at different times for Mn-Al-Fe alloy. (**c**) UV absorption spectra at different times for Mn-Al-Co alloy. (**d**) The decolorization % from UV absorption intensity at 597 nm versus reaction time for both alloys.

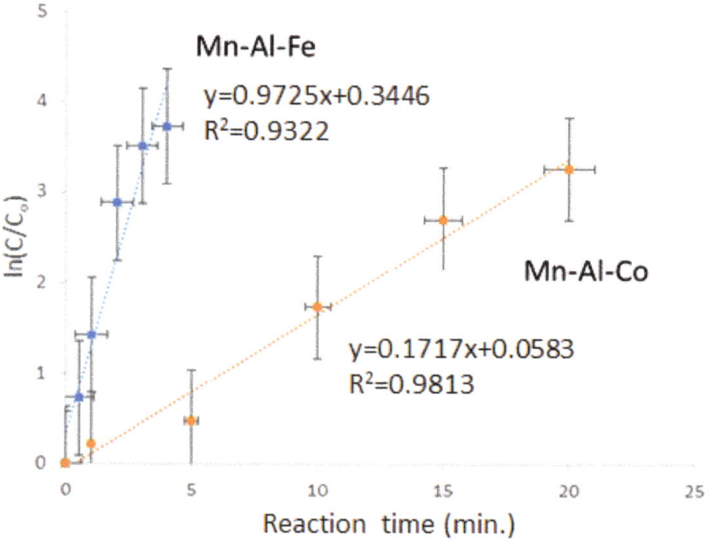

Figure 4. Ln(C/C_o) versus reaction time of the MA Mn-Al-Fe and Mn-Al-Co alloys.

The appearance of bubbles was detected inside the solution. This effect is associated with the formation of hydrogen gas. Resting upon this fact, the proposed mechanism of the dye degradation operates as a result of the release of: (a) H_2 from the reduction in water, (b) the cleavage of -N=N- bonds and, (c), the formation of $-NH_2$ groups. Under these conditions, the release of hydrogen gas is accompanied by the formation of insoluble metal hydroxides [19,37]. The mechanism for Mn-Al-Fe and Mn-Al-Co, in acid conditions, are based on the reactions of these references. Here, we introduce Fe and Co reactions. The reactions involving Fe are:

$$Fe \rightarrow Fe^{3+} + 3e^- \quad (1)$$

$$Fe^{3+} + 3(OH)^- \rightarrow Fe(OH)_3 \quad (2)$$

The main difference with previous works in Mn-Al alloys is the role of Fe. The mechanism for Mn-Al-Co is similar, taking into account the Co reactions ($pH < 7$) in substitution of Fe reactions [38]

$$Co \rightarrow Co^{2+} + 2e^- \quad (3)$$

$$Co^{2+} + 2(OH)^- \rightarrow Co(OH)_2 \quad (4)$$

The aluminum oxide can act as an acid or a basic agent relying on the environment by the Al $(OH)_3$ and [Al $(OH)_4]^-$ formation. The pH of the solution increases (from 6.3 to 10.8 or 11.2 in Mn-Al-Fe/Mn-Al-Co, respectively) due to OH^- ion enrichment. The mechanism proposed by the Mn-Al-Fe and Mn-Al-Co reaction, in basic conditions, ($pH > 7$), is given in references [19,38]. The reactions involving Fe and Co are the same.

To assess the product of the discoloration reaction by the Mn-Al-Fe (or Mn-Al-Co) powder, we observed the surface of the particles by scanning electron microscopy. The micrographs obtained are plotted in Figures 5 and 6 and unveil that some products of the reaction are spread (uniformly distributed) over the entire particle's surface. The same effect was found in other metallic compounds [39]. These crystalline precipitates are mainly composed of the element Al and oxygen. In addition, a high amount of corrosion holes was detected on the surface of the alloys, indicating that corrosion of the alloy particles during the degradation process of RB5 occurs by pitting [19]. It is well known that the corrosion of the Al-based alloy is dependent on the damage of the Al_2O_3-passive layer. In addition, it is believed that pitting corrosion is one of the main degradation mechanisms of RB5. This result is very expressive, the addition of an acid to the initially alkaline textile industrial wastewater unnecessary. Likewise, the cationic metals concentrations of Mn, Al, Fe and Co are obtained from absorption spectrometry: 2.12, <0.59, <0.16 and <0.20 mg L^{-1} (detection limit) for Mn, Al, Fe and Co, respectively. The dissolution of Al $(OH)_3$ results from the excess of OH^-, and therefore more surface is exposed to water favoring the progression of the reactions.

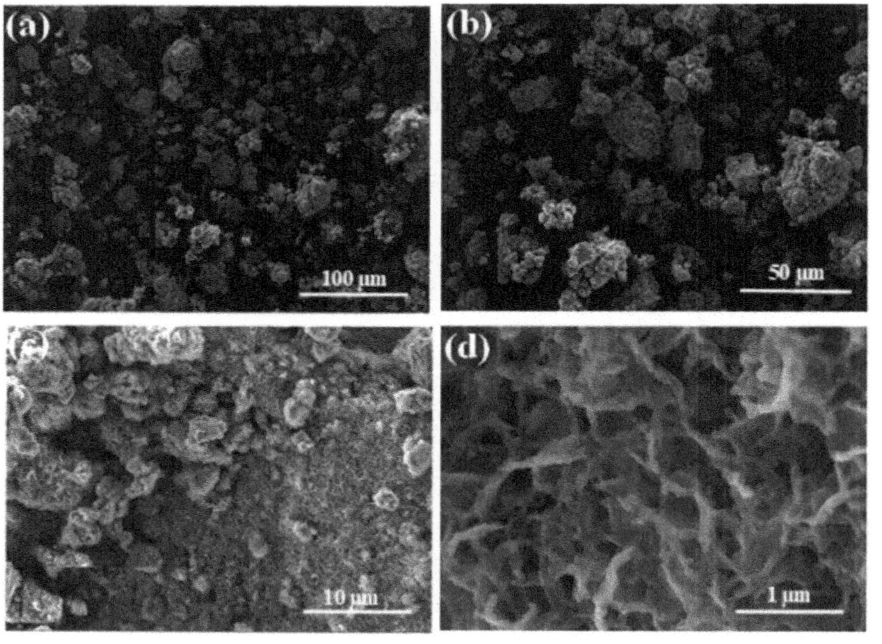

Figure 5. SEM micrographs of MA Mn-Al-Fe after RB5 degradation at different magnifications. Scale bars: (**a**) 100 μm, (**b**) 500 μm, (**c**) 10 μm and (**d**) 1 μm.

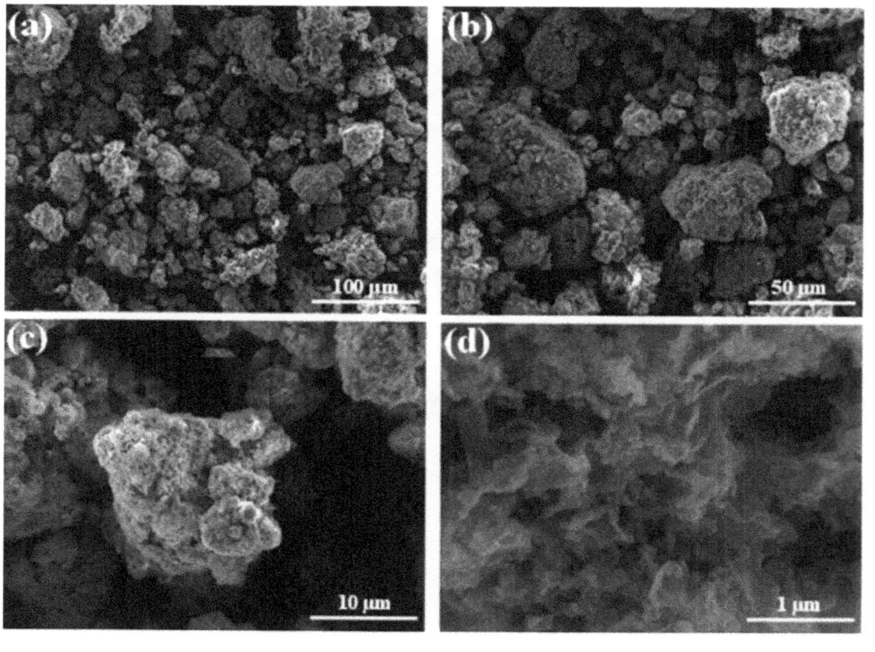

Figure 6. SEM micrographs of MA Mn-Al-Co after RB5 degradation at different magnifications. Scale bars: (**a**) 100 μm, (**b**) 500 μm, (**c**) 10 μm and (**d**) 1 μm.

After the decoloring reaction, the size distributions of the Mn-Al-Fe and Mn-Al-Co alloys are illustrated in Figure 7a,c, respectively, showing mean sizes of the order of 20 and 17 μm. This difference in size is associated to the precipitates on the surface of the nanostructured powders. The corresponding EDS analyzes depicted in Figure 7b,d indicate the approximately compositions of the elements, Mn 50.0/Al: 15.0/Fe: 8.0/O: 27.0 for the Mn-Al-Fe alloy, and Mn: 47.0/Al: 17.0/Co: 10.0/O: 26.0 for the Mn-Al-Co alloy. Finally, the microstructural study of Mn-Al-Fe and Mn-Al-Co powders before and after RB5 dye degradation (pH < 7) is performed by XRD. The results are shown (see Figure 8a,b). The diffraction patterns obtained before the degradation of the two Mn-Al-Fe/Mn-Al-Co powders reveal the coexistence of a solid solution of $MnAlFe_2$ (JCPD 00-054-0388) and $MnAlCo_2$ (JCPD 03-065-5185) phases, respectively, (B2, space group Fm3m) with traces of very small proportions of MnAl phase (η) rich in Al and tetragonal face-centered (FCT) structure [40]. The average crystallite size of the solid solution B2 is of 60 ± 3 nm. This nanocrystalline structure may stand for another factor increasing the reactivity of these metal powders within the aqueous dyed solutions by increasing atoms in the boundary regions. As two phases are detected after MA, the samples are not chemically homogeneous. Nevertheless, the fast discoloration kinetics allow us to state that this chemical in-homogeneity is probably not determinant. The inspection of the X-ray diffraction patterns of both powdered alloys obtained after the discoloration reaction demonstrates the appearance of new phases identified as: $Al(OH)_3$ (JCPD 00-003 -0915), $Fe(OH)_3$ (JCPD 00-038-0032), and $Co(OH)_2$ (JCPD 00-051-173) beside the B2-MnAl (Fe, Co)$_2$ and η-MnAl phases (Figure 8a,b). The products resemble micro-precipitates of $Al(OH)_3$, $Fe(OH)_3$ and $Co(OH)_2$ hydroxides, covering the surface of the particles. These hydroxides are minor phases (>10 at.%). It should be remarked that it is likely that in all phases, including hydroxides, all elements (Mn, Al, Fe or Co) are present: one with a high amount and the other two elements in solid solution. Likewise, EDS analysis gives information about the volume of the powder near the surface, not the overall volume of the powders.

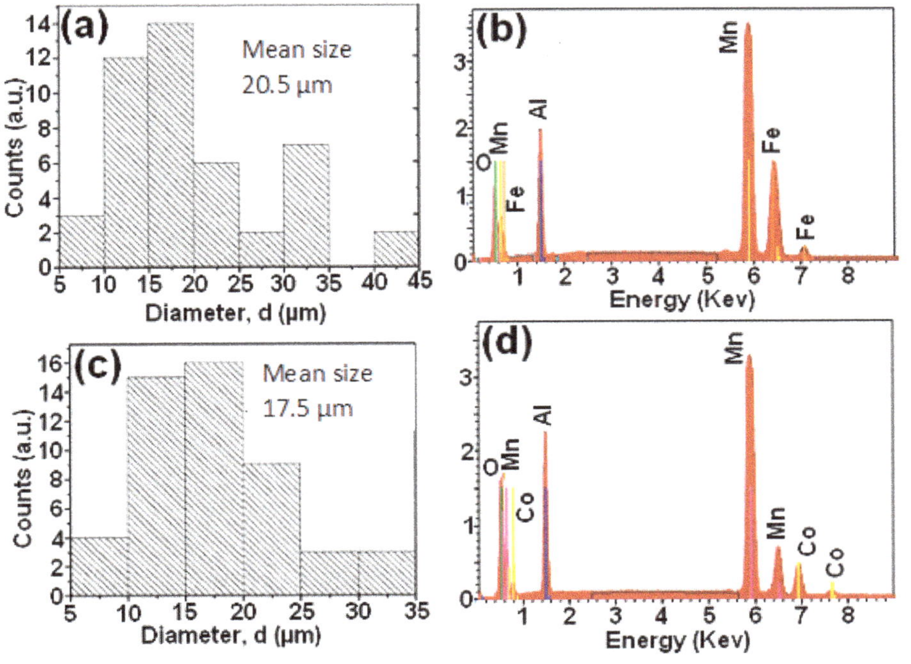

Figure 7. Distribution of particle sizes of MA powders after decoloring reaction: (**a**) Mn-Al-Fe, (**b**) Mn-Al-Co. EDS microanalysis: (**c**) Mn-Al-Fe and (**d**) Mn-Al-Co.

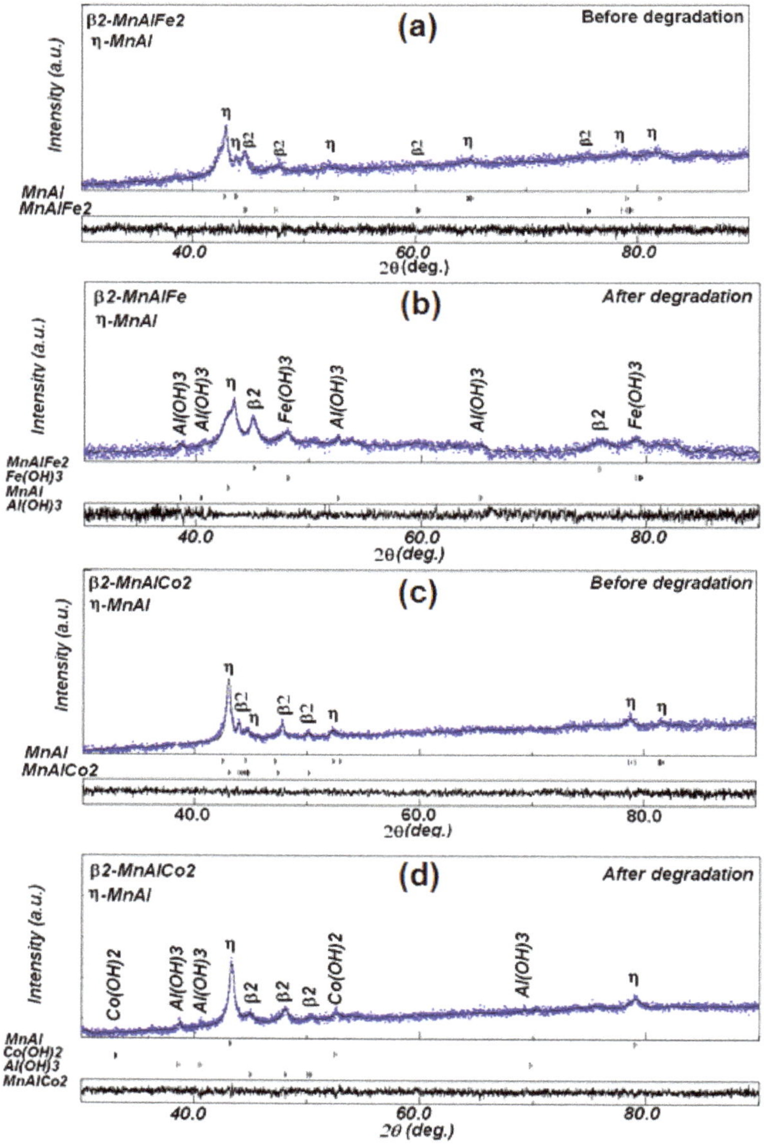

Figure 8. XRD diffraction patterns: (**a**,**b**) MA Mn-Al-Fe powder before and after degradation and (**c**,**d**) MA Mn-Al-Co powder before and after degradation.

It has been reported that MA can favor a higher specific surface area [22]. This parameter, determined by BET, provides best information about the active surface than the statistical analysis of powders' size distribution from SEM micrographs. The higher the specific surface, the higher the interaction between particles and the dye molecule. The specific surface area was 0.50 ± 0.02 $m^2\ g^{-1}$ for the Mn-Al-Fe alloy and 0.45 ± 0.02 $m^2\ g^{-1}$ for the Mn-Al-Co alloy. There is apparently a slight difference in the specific surface. Nevertheless, fast discoloration was found in the sample with Fe addition (the slopes of Figure 4 are 0.9725 and 0.1727 s^{-1} in alloys with Fe and Co, respectively). This high difference is due to the high surface area of the sample with Fe, but can also be partially attributed, as discussed above, to the different chemical compositions of the alloys.

Thus, the results make these two alloys promising candidates for the fast degradation of different azo dyes. In Mn-Al MA powders, the specific surface values were found to range between 0.55 and 0.48 $m^2\ g^{-1}$. Values between 0.50 and 0.55 $m^2\ g^{-1}$ were found by other authors [41].

The FTIR spectra of RB5 before and after degradation were shown in Figure 9. The main peaks and bands identification are given in Table 1.

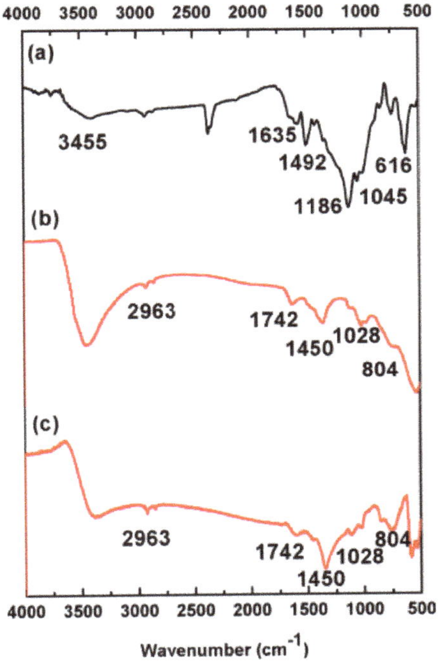

Figure 9. FTIR spectra: (**a**) RB5 powder before degradation selection. (**b**,**c**) Mn-Al-Fe and Mn-Al-Co powders after degradation, respectively.

Table 1. FTIR peaks/bands identification [42–46].

Peak/Band	Identification
3455 cm^{-1}	O-H stretching vibration
1742 cm^{-1}	C=C stretching vibration
1635 cm^{-1}	azo bond (-N=N-)
1528 cm^{-1}	N-H stretching vibration
1492 cm^{-1}	C= C aromatic skeletal vibrations
1260 cm^{-1}	C-N stretching vibration
1186 cm^{-1}	C-OH stretching vibration
1045 cm^{-1}	C-OH stretching vibration
1028 cm^{-1}	benzene mode coupling with stretching vibration of $-SO_3$
804 cm^{-1}	$-CH_3$ skeletal vibration
616 cm^{-1}	sulfonic group

From FTIR spectra, it is clear that some characteristic RB5 peaks decrease during the reaction, while some new peaks appear. The decreasing peaks are the azo bond (-N=N-, at 1635 cm^{-1}) and the naphthalene ring (C=C aromatic skeletal vibration, at 1400–1600 cm^{-1}). A minor peak remains in all FTIR spectra, the 2963 cm^{-1} peak, assigned to the skeletal vibration of the benzene ring. Some new peaks appear after degradation, located at 1742, 1528 and 1260 cm^{-1} (stretching vibration of C=C, N-H and C-N bonds, respectively). This demonstrates that the reductive cleavage of the -N=N- bond [19,44] favors the presence of amines. It should be remarked that amines are also water-pollutant. Mn-Al based alloys' application in industrial processes will be combined with an ulterior amine adsorption process. Thus, metallic nanostructured particles are candidates for the first step of a combined multi-step procedure.

In addition, these results are in good accordance with the previously stated UV-Vis (Figure 3b,c) regarding the cleavage of the azo bands [42,44–46]. The FTIR confirms that the results indicated that the azo bond and the naphthalene ring are broken by mechanical alloying of Mn-Al-(Fe, Co) and result in the formation of amino and alkyl compounds.

The HPLC chromatograms are shown in Figure 10. Before degradation, the major peak has a molecular weight of 991. Similar results were elaborated by Shilpa and coworkers [46], Patel and coworkers [47] and Ben Mbarek [19,20]. The RB5 chromatography should be compared with the HPLC chromatography and LC-MS mass spectrums of the solutions after degradation with the metallic alloys.

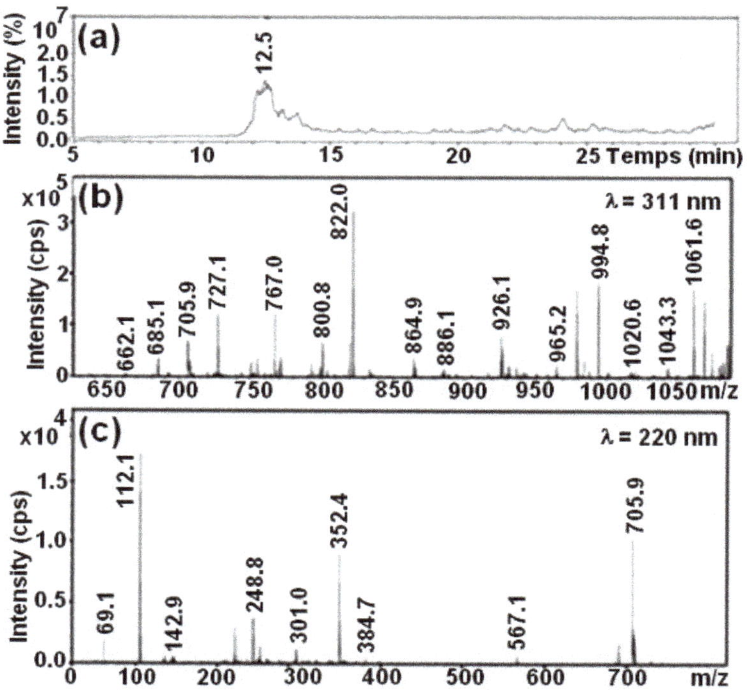

Figure 10. (**a**) HPLC profile of RB5 dye solution. Mass spectrums of peaks from 10.5 min at (**b**) λ = 311 nm and (**c**) λ = 220 nm.

Figure 11 displays the LC-MS analyses of the extract of the two solutions obtained following the degradation reactions carried out by the two ternary powders Mn-Al-Fe and Mn-Al-Co. These analyses reveal several retention time peaks. In addition, analyses of the major peaks eluted between 10 and 15 min demonstrated the existence of two peaks at 349 and 280.1 m/z in relation to the final amino molecules, related to 1-2-7-triamino-8-hydroxy,3-6-naphthalenedisulfonate and 1-sulfonic acid, 2-(4-aminobenzenesulfonyl) ethanol, as depicted in Figure 10a,b. All these results confirm the appearance of aromatic amino compounds emerging from the decomposition reaction of the large RB5 organic molecule. This is inferred through the appearance of signals at 248.9 m/z, related to the aromatic amines obtained after bond cleavage. The mechanism of discoloration of RB5 with these metal powders can only be identical to that identified for the case of the Mn-Al binary following the cleavage of the azo bonds [19].

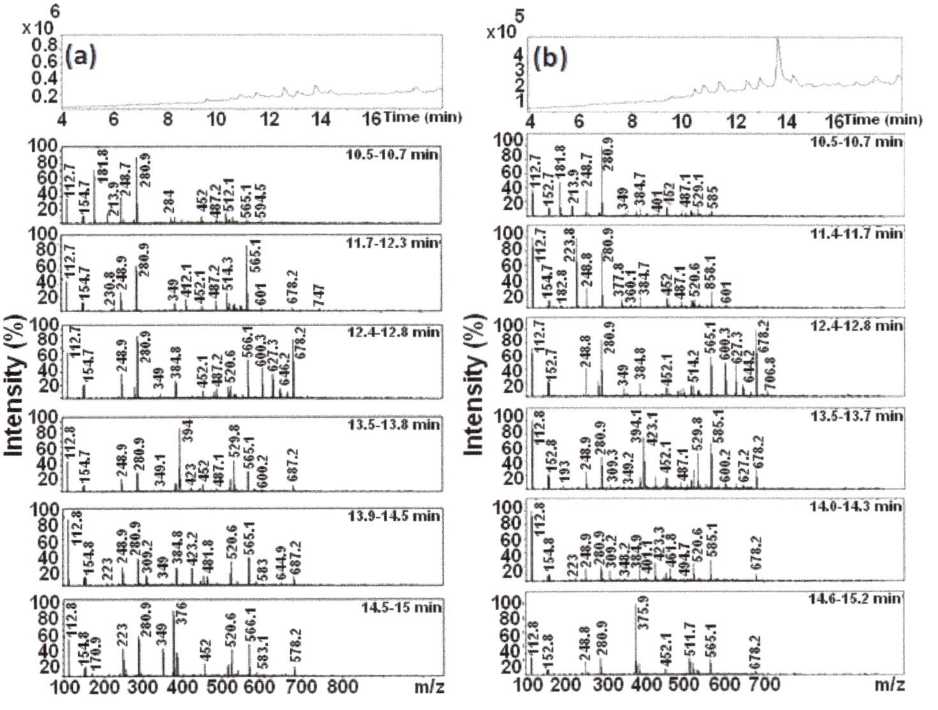

Figure 11. HPLC evolution profiles after degradation of the RB5 dye and mass spectrum of peaks from 10.5 to 15.2 min of the MA powders: (**a**) Mn-Al-Fe and (**b**) Mn-Al-Co.

4. Conclusions

In sum, Mn-Al-Fe and Mn-Al-Co powders were successfully manufactured by mechanical alloying. The efficiency and kinetics of the discoloration reaction of aqueous solutions of RB5 are systematically evaluated to determine the capacity of these two alloys in the bleaching reaction.

- The Mn-Al-Fe powder displays an excellent degradation efficiency and the kinetics of the reaction are proven to be faster than that of Mn-Al based alloys with 10 at.% of Fe and Co. The high efficiency of the Mn-Al-Fe powder is associated with the configuration of valence electrons. This favors a higher concentration of reactive (hole) sites in the d-band for iron than that for cobalt (forming a localized adsorption bond with the adsorbate molecule), favoring a stronger adsorption capacity;
- For the ternary alloy with Fe, the solution is fully discolored in 5 min, whereas for the solution with Co, the reaction time to full decolorization is 20 min. The reductive cleavage of the -N=N- bond was demonstrated by FTIR spectra analysis;
- Furthermore, the zero-valent iron (Fe^0) and the zero valent cobalt (Co^0) become reaction actors to promote the production of hydrogen from water. Likewise, the decolorization of R5 solution is

consistent with first-order kinetics. Hence, the high efficiency of Mn-Al based alloys recorded in decolorization treatments of dyed wastewaters, already emphasized in previous works, can be modified and increased by minor alloying with other transition metals.

Thus, these promising materials have interesting characteristics to be applied, probably in a multi-step procedure in the wastewater treatment in the textile industry.

Author Contributions: Methodology, L.E., J.S.; formal analysis, W.B.M.; investigation, W.B.M.; resources, E.P., J.-J.S.; data curation, L.E., M.K.; writing—original draft preparation, W.B.M., J.-J.S.; supervision, L.E.; funding acquisition, E.P., J.-J.S. All authors have read and agreed to the published version of the manuscript.

Funding: Work funded by Spanish MINECO Grant No. FIS2017-82625-P, Catalan GenCat Grant No. 2017SGR0042 and University of Girona UNIGE-2-2019 project is also acknowledged.

Acknowledgments: The authors acknowledge the technical facilities of the STR support to research units of the University of Girona.

Conflicts of Interest: The authors declare no conflict of interest.

References

1. Amin, N.K. Removal of direct blue-106 dye from aqueous solution using new activated carbons developed from pomegranate peel: Adsorption equilibrium and kinetics. *J. Hazard. Mater.* **2009**, *165*, 52–62. [CrossRef] [PubMed]
2. Fu, W.; Yang, H.; Chang, L.; Hari-Bala; Li, M.; Zou, G. Anatase TiO2 nanolayer coating on strontium ferrite nanoparticles for magnetic photocatalyst. *Colloids Surf. A Physicochem. Eng. Asp.* **2006**, *289*, 47–52. [CrossRef]
3. Daneshvar, N.; Aber, S.; Khani, A.; Khataee, A.R. Study of imidaclopride removal from aqueous solution by adsorption onto granular activated carbon using an on-line spectrophotometric analysis system. *J. Hazard. Mater.* **2007**, *144*, 47–51. [CrossRef] [PubMed]
4. Walker, S.; Narbaitz, R.M. Hollow fiber ultrafiltration of Ottawa River water: Floatation versus sedimentation pre-treatment. *Chem. Eng. J.* **2016**, *288*, 228–237. [CrossRef]
5. Okolo, B.I.; Nnaji, P.C.; Onukwuli, O.D. Nephelometric approach to study coagulation-flocculation of brewery effluent medium using Detarium microcarpum seed powder by response surface methodology. *J. Environ. Chem. Eng.* **2016**, *4*, 992–1001. [CrossRef]
6. Gupta, M.K.; Mittal, A.K. Integrated biological and advanced oxidation based treatment of hexamine bearing wastewater: Effect of cow-dung as a co-substrate. *J. Hazard. Mater.* **2016**, *308*, 394–401. [CrossRef]
7. Khani, A.; Sohrabi, M.R.; Khosravi, M.; Davallo, M. Enhancing purification of an azo dye solution in nanosized zero-valent iron-ZnO photocatalyst system using subsequent semibatch packed-bed reactor. *Turkish J. Eng. Environ. Sci.* **2013**, *37*, 91–99. [CrossRef]
8. Khani, A.; Pezeshki, B. Easy simultaneous synthesis–immobilization of nanosized CuO–ZnO on perlite as a photocatalyst for degradation of acid orange 7 from aqueous solution in the presence of visible light. *Desalin. Water Treat.* **2016**, *57*, 7047–7053. [CrossRef]
9. Xie, T.; Yang, J.; Peng, Y.; Wang, J.; Liu, S.; Xu, L.; Liu, C. β-Bi2O3/SrFe12O19 magnetic photocatalyst: Facile synthesis and its photocatalytic activity. *Mater. Technol.* **2019**, *34*, 843–850. [CrossRef]
10. Kanel, S.R.; Manning, B.; Charlet, L.; Choi, H. Removal of arsenic(III) from groundwater by nanoscale zero-valent iron. *Environ. Sci. Technol.* **2005**, *39*, 1291–1298. [CrossRef]
11. Schrick, B.; Blough, J.L.; Jones, A.D.; Mallouk, T.E. Hydrodechlorination of trichloroethylene to hydrocarbons using bimetallic nickel-iron nanoparticles. *Chem. Mater.* **2002**, *14*, 5140–5147. [CrossRef]
12. Chang, J.H.; Cheng, S.F. The remediation performance of a specific electrokinetics integrated with zero-valent metals for perchloroethylene contaminated soils. *J. Hazard. Mater.* **2006**, *131*, 153–162. [CrossRef] [PubMed]
13. Xiong, Z.; Zhao, D.; Pan, G. Rapid and complete destruction of perchlorate in water and ion-exchange brine using stabilized zero-valent iron nanoparticles. *Water Res.* **2007**, *41*, 3497–3505. [CrossRef] [PubMed]
14. Hu, J.; Lo, I.M.C.; Chen, G. Fast removal and recovery of Cr(VI) using surface-modified jacobsite (MnFe2O4) nanoparticles. *Langmuir* **2005**, *21*, 11173–11179. [CrossRef] [PubMed]
15. ChangQin, Z.; ZhengWang, Z.; HaiFeng, Z.; ZhuangQi, H. Article Rapid reductive degradation of azo dyes by a unique structure of amorphous alloys. *Chin. Sci. Bull.* **2011**, *56*, 3988–3992. [CrossRef]

16. Zhang, Z.; Wang, Y.; Qi, Z.; Zhang, W.; Qin, J.; Frenzel, J. Generalized fabrication of nanoporous metals (Au, Pd, Pt, Ag, and Cu) through chemical dealloying. *J. Phys. Chem. C* **2009**, *113*, 12629–12636. [CrossRef]
17. Pugh, D.V.; Dursun, A.; Corcoran, S.G. Electrochemical and morphological characterization of Pt-Cu dealloying. *J. Electrochem. Soc.* **2005**, *152*, B455–B459. [CrossRef]
18. Mangipudi, K.R.; Epler, E.; Volkert, C.A. Morphological similarity and structure-dependent scaling laws of nanoporous gold from different synthesis methods. *Acta Mater.* **2017**, *140*, 337–343. [CrossRef]
19. Ben Mbarek, W.; Azabou, M.; Pineda, E.; Fiol, N.; Escoda, L.; Suñol, J.J.; Khitouni, M. Rapid degradation of azo-dye using Mn-Al powders produced by ball-milling. *RSC Adv.* **2017**, *7*, 12620–12628. [CrossRef]
20. Ben Mbarek, W.; Pineda, E.; Escoda, L.; Suñol, J.J.; Khitouni, M. High efficiency decolorization of azo dye Reactive Black 5 by Ca-Al particles. *J. Environ. Chem. Eng.* **2017**, *5*, 6107–6113. [CrossRef]
21. AboliGhasemabadi, M.; Mbarek, W.B.; Casabella, O.; Roca-Bisbe, H.; Pineda, E.; Escoda, L.; Suñol, J.J. Application of mechanically alloyed MnAl particles to de-colorization of azo dyes. *J. Alloys Compd.* **2018**, *741*, 240–245. [CrossRef]
22. AboliGhasemabadi, M.; Ben Mbarek, W.; Cerrillo-Gil, A.; Roca-Bisbe, H.; Casabella, O.; Blánquez, P.; Pineda, E.; Escoda, L.; Suñol, J.J. Azo-dye degradation by Mn–Al powders. *J. Environ. Manag.* **2020**, *258*, 110012. [CrossRef] [PubMed]
23. Petcharoen, K.; Sirivat, A. Synthesis and characterization of magnetite nanoparticles via the chemical co-precipitation method. *Mater. Sci. Eng. B Solid State Mater. Adv. Technol.* **2012**, *177*, 421–427. [CrossRef]
24. Devatha, C.P.; Thalla, A.K.; Katte, S.Y. Green synthesis of iron nanoparticles using different leaf extracts for treatment of domestic waste water. *J. Clean. Prod.* **2016**, *139*, 1425–1435. [CrossRef]
25. Wu, S.; Pan, Y.; Lu, J.; Wang, N.; Dai, W.; Lu, T. Effect of the addition of Mg, Ti, Ni on the decoloration performance of AlCrFeMn high entropy alloy. *J. Mater. Sci. Technol.* **2019**, *35*, 1629–1635. [CrossRef]
26. Feng, Y.P.; Gaztelumendi, N.; Fornell, J.; Zhang, H.Y.; Solsona, P.; Baró, M.D.; Suriñach, S.; Ibáñez, E.; Barrios, L.; Pellicer, E.; et al. Mechanical properties, corrosion performance and cell viability studies on newly developed porous Fe-Mn-Si-Pd alloys. *J. Alloys Compd.* **2017**, *724*, 1046–1056. [CrossRef]
27. Hermawan, H.; Alamdari, H.; Mantovani, D.; Dubé, D. Iron-manganese: New class of metallic degradable biomaterials prepared by powder metallurgy. *Powder Metall.* **2008**, *51*, 38–45. [CrossRef]
28. Cao, J.; Wei, L.; Huang, Q.; Wang, L.; Han, S. Reducing degradation of azo dye by zero-valent iron in aqueous solution. *Chemosphere* **1999**, *38*, 565–571. [CrossRef]
29. Feng, W.; Nansheng, D.; Helin, H. Degradation mechanism of azo dye C. I. reactive red 2 by iron powder reduction and photooxidation in aqueous solutions. *Chemosphere* **2000**, *41*, 1233–1238. [CrossRef]
30. Stylidi, M.; Kondarides, D.I.; Verykios, X.E. Pathways of solar light-induced photocatalytic degradation of azo dyes in aqueous TiO2 suspensions. *Appl. Catal. B Environ.* **2003**, *40*, 271–286. [CrossRef]
31. Zhang, C.; Zhang, H.; Lv, M.; Hu, Z. Decolorization of azo dye solution by Fe-Mo-Si-B amorphous alloy. *J. Non. Cryst. Solids* **2010**, *356*, 1703–1706. [CrossRef]
32. Zhang, C.; Zhu, Z.; Zhang, H. Effects of the addition of Co, Ni or Cr on the decolorization properties of Fe-Si-B amorphous alloys. *J. Phys. Chem. Solids* **2017**, *110*, 152–160. [CrossRef]
33. Wang, J.Q.; Liu, Y.H.; Chen, M.W.; Xie, G.Q.; Louzguine-Luzgin, D.V.; Inoue, A.; Perepezko, J.H. Rapid degradation of Azo dye by Fe-based metallic glass powder. *Adv. Funct. Mater.* **2012**, *22*, 2567–2570. [CrossRef]
34. Nam, S.; Tratnyek, P.G. Reduction of azo dyes with zero-valent iron. *Water Res.* **2000**, *34*, 1837–1845. [CrossRef]
35. Zhang, C.; Zhu, Z.; Zhang, H.; Hu, Z. Rapid decolorization of Acid Orange II aqueous solution by amorphous zero-valent iron. *J. Environ. Sci.* **2012**, *24*, 1021–1026. [CrossRef]
36. Zhang, C.; Zhu, Z.; Zhang, H.; Sun, Q.; Liu, K. Effects of cobalt content on the decolorization properties of Fe-Si-B amorphous alloys. *Results Phys.* **2018**, *10*, 1–4. [CrossRef]
37. Mook, W.T.; Aroua, M.K.; Szlachta, M.; Lee, C.S. Optimisation of Reactive Black 5 dye removal by electrocoagulation process using response surface methodology. *Water Sci. Tech.* **2017**, *75*, 852. [CrossRef]
38. Garcia, E.M.; Santos, J.S.; Pereira, E.C.; Freitas, M.B.J.G. Electrodeposition of cobalt from spent Li-ion battery cathodes by the electrochemistry quartz crystal microbalance technique. *J. Power Sources* **2008**, *185*, 549–553. [CrossRef]

39. Zhang, L.; Gao, X.; Zhang, Z.; Zhang, M.; Cheng, Y.; Su, J. A doping lattice of aluminum and copper with accelerated electron transfer process and enhanced reductive degradation performance. *Sci. Rep.* **2016**, *6*, 31797. [CrossRef]
40. Paduani, C.; Schaf, J.; Persiano, A.I.C.; Ardisson, J.D.; Takeuchi, A.Y.; Riegel, I.C. Strong dependence of ferrimagnetic properties on Co concentration in the Mn1-xAl1-yCox+y system. *Intermetallics* **2010**, *18*, 1659–1662. [CrossRef]
41. Stylidi, M.; Kondarides, D.I.; Verykios, X.E. Visible light-induced photocatalytic degradation of Acid Orange 7 in aqueous TiO2 suspensions. *Appl. Catal. B Environ.* **2004**, *47*, 189–201. [CrossRef]
42. Méndez-Martínez, A.J.; Dávila-Jiménez, M.M.; Ornelas-Dávila, O.; Elizalde-González, M.P.; Arroyo-Abad, U.; Sirés, I.; Brillas, E. Electrochemical reduction and oxidation pathways for Reactive Black 5 dye using nickel electrodes in divided and undivided cells. *Electrochim. Acta* **2012**, *59*, 140–149. [CrossRef]
43. Almeida, E.J.R.; Corso, C.R. Comparative study of toxicity of azo dye Procion Red MX-5B following biosorption and biodegradation treatments with the fungi Aspergillus niger and Aspergillus terreus. *Chemosphere* **2014**, *112*, 317–322. [CrossRef] [PubMed]
44. Elías, V.R.; Sabre, E.V.; Winkler, E.L.; Satuf, M.L.; Rodriguez-Castellón, E.; Casuscelli, S.G.; Eimer, G.A. Chromium and titanium/chromium-containing MCM-41 mesoporous silicates as promising catalysts for the photobleaching of azo dyes in aqueous suspensions. A multitechnique investigation. *Microporous Mesoporous Mater.* **2012**, *163*, 85–95. [CrossRef]
45. Agrawal, S.; Tipre, D.; Patel, B.; Dave, S. Optimization of triazo Acid Black 210 dye degradation by Providencia sp. SRS82 and elucidation of degradation pathway. *Process Biochem.* **2014**, *49*, 110–119. [CrossRef]
46. Shilpa, S.; Shikha, R. Biodegradation of Dye Reactive Black-5 by a Novel Bacterial Endophyte. *Int. Res. J. Environ. Sci.* **2015**, *4*, 44–53. [CrossRef]
47. Patel, R.; Suresh, S. Decolourization of azo dyes using magnesium-palladium system. *J. Hazard. Mater.* **2006**, *137*, 1729–1741. [CrossRef]

metals

Article

Structure-Phase Transformations in the Course of Solid-State Mechanical Alloying of High-Nitrogen Chromium-Manganese Steels

Kirill Lyashkov [1], Valery Shabashov [2,*], Andrey Zamatovskii [2], Kirill Kozlov [2], Natalya Kataeva [2], Evgenii Novikov [2] and Yurii Ustyugov [2]

1 Institute of Engineering Science, Ural Branch, Russian Academy of Sciences, 620108 Ekaterinburg, Russia; lyashkov@imp.uran.ru

2 Mikheev Institute of Metal Physics, Ural Branch, Russian Academy of Sciences, 620108 Ekaterinburg, Russia; zamatovsky@imp.uran.ru (A.Z.); kozlov@imp.uran.ru (K.K.); kataeva@imp.uran.ru (N.K.); evg_nov@mail.ru (E.N.); ustyugov@imp.uran.ru (Y.U.)

* Correspondence: shabashov@imp.uran.ru

Abstract: The solid-state mechanical alloying (MA) of high-nitrogen chromium-manganese austenite steel—MA in a planetary ball mill, —was studied by methods of Mössbauer spectroscopy and transmission electron microscopy (TEM). In the capacity of a material for the alloying we used mixtures of the binary Fe–Mn and Fe–Cr alloys with the nitrides CrN (Cr_2N) and Mn_2N. It is shown that ball milling of the mixtures has led to the occurrence of the $\alpha \rightarrow \gamma$ transitions being accompanied by the (i) formation of the solid solutions supersaturated with nitrogen and by (ii) their decomposition with the formation of secondary nitrides. The austenite formed by the ball milling and subsequent annealing at 700–800 °C, was a submicrocrystalline one that contained secondary nano-sized crystalline CrN (Cr_2N) nitrides. It has been established that using the nitride Mn_2N as nitrogen-containing addition is more preferable for the formation and stabilization of austenite—in the course of the MA and subsequent annealing—because of the formation of the concentration-inhomogeneous regions of γ phase enriched with austenite-forming low-mobile manganese.

Keywords: austenitic alloys; mechanical alloying; high-nitrogen steels; atomic redistribution; point defects; Mössbauer spectroscopy

Citation: Lyashkov, K.; Shabashov, V.; Zamatovskii, A.; Kozlov, K.; Kataeva, N.; Novikov, E.; Ustyugov, Y. Structure-Phase Transformations in the Course of Solid-State Mechanical Alloying of High-Nitrogen Chromium-Manganese Steels. *Metals* **2021**, *11*, 301. https://doi.org/10.3390/met11020301

Academic Editors: Joan-Josep Suñol and Carlos Garcia-Mateo

Received: 16 December 2020
Accepted: 4 February 2021
Published: 9 February 2021

1. Introduction

In the modern technology of production of the materials with specific (improved) functional properties a great emphasis is placed on the design of the high-nitrogen steels, alloys, and composites [1–3]. This is due to the (i) need for economically alloyed steels and (ii) feasibility of substitution of nitrogen for the high-cost nickel and manganese. The current technology for melting high-nitrogen steels under nitrogen pressure requires expensive equipment and high energy expenditure. In the capacity of an alternative and cheaper method for producing high-nitrogen steels, today powder metallurgy methods using mechanical alloying are proposed. In most studies on the mechanical alloying (MA) of high-nitrogen steels, the metal powder was saturated with nitrogen from the gas phase in a flowing atmosphere of nitrogen or ammonia [4–13]. In these works, the authors usually employed vibrationally assisted ball milling with long-term processing times (100 h or more). In the works [14–20], for the purpose of nitriding Fe-based steels, the authors were the first to have employed the method of solid-phase MA (i.e., MA of materials in a solid state). This method does not require additional gas equipment to create a flowing nitrogen-containing atmosphere in the mill vessel. In addition, the method does not require such a long time of mechanical processing and allows for one to use planetary-type ball mills. To the basis of the approach in [15–18] their authors have placed the cyclically evaluating deformation-induced phase transitions «dissolution—precipitation» of disperse particles of the prime low-stable nitrides (of the types CrN and Fe_4N) located in metallic matrices. In result of the deformation-induced dissolution of prime (initially present) nitrides, in the metallic matrix of alloys one can observe the formation of both the nano-structured solid solutions supersaturated with nitrogen and highly disperse secondary nitrides. Within the framework of the proposed approach with the employment of the (i) method of severe plastic deformation (namely, by high pressure torsion in the rotating Bridgman anvils) [15,16], (ii) friction-induced external action [16], and (iii) ball milling (BM) [17–20], it became feasible to have performed alloying of the Fe–Ni–Cr–N and the Fe–Mn–Cr–N austenite. For the development of high-nitrogen steels as competitors to austenite Fe–Ni–Cr–N alloys, the alloys based on the system Fe–Cr–Mn–N were proposed in [21,22] for consideration. First experiments on the solid-state MA of Fe–Cr–Mn–N steels using BM were performed on the mixtures of the binary Fe–Mn alloy and CrN (Cr_2N) nitrides [17], as well as on the mixtures of the pure metals Fe, Mn, and Cr—on the one hand, and the nitrides CrN and Mn_2N, on the other [19].

The purpose of this work is the investigation of the possibility to obtaining nitrogen-containing austenite by means of employment of the solid-phase MA where in the capacity of the starting material a researcher uses initial mixtures of the binary alloys and nitrides together with the alloying elements that traditionally are taken as constituents of the stainless steels from the austenite class of Fe–Cr–Mn–N type. The subject of the study was the analysis of the mechanism, kinetics, and properties of the products of the MA—in the course of BM—and subsequent thermal anneals, in dependence of initial compositions of the initial matrices and nitrides. In particular, an optimum composition of the austenite to obtain was of interest from the point of view of reducing the volume of alloying manganese and chromium in the composition of MA-produced austenite.

2. Experimental

2.1. Sample Compositions and Their Treatment

For obtaining MA Fe–Cr–Mn–N alloys, in the capacity of their metallic matrix we used powders of the binary alloys with BCC crystal lattice of compositions: Fe–*X*Mn (*X*, wt% = 0, 4.0, 6.7, 8.9) and Fe–*Y*Cr (*Y*, wt% = 0, 4.7, 8.6, 14.2), and in the capacity of "donor" of the alloying element «nitrogen» there were chosen the nitrides CrN (Cr_2N) and Mn_2N, which are low-stable in conditions of deformation. We analyzed the mixtures in the series *A* and *B* (see Tables 1 and 2), each of which had several compositions of constituents with the varying contents of the metallic matrix and 10 or 20 wt% proportion of the nitrogen-containing addition:

$$A\text{: Fe-}X\text{Mn} + 20\text{CrN } (\text{Cr}_2\text{N});$$

$$B\text{: Fe-}Y\text{Cr} + (10 \text{ or } 20\text{Mn}_2\text{N}).$$

Table 1. The formula of composition, the content of alloying elements, and the quantity of austenite in the mechanical alloying (MA) alloys of series *A* after the ball milling and subsequent anneals.

No.	Formula of Composition	Ball Milling for 10 h			Annealing after BM at 700 °C (*a*) and 800 °C (*b*)					
	FeXMn + 20CrN (*X* in wt%)	Quantity of Austenite γ, vol.%	Content of Nitrogen in Austenite γC_N, wt%	Contents of Cr and Mn in the α Phase αC_{Mn+Cr}, wt%	Quantity of Austenite γ, vol.%		Content of Nitrogen in Austenite γC_N, wt%		Contents of Cr and Mn in the α Phase αC_{Mn+Cr}, wt%	
					a	*b*	*a*	*b*	*a*	*b*
1	X = 0.0	6	3.7	4.0						
2	X = 4.0	7	1.4	5.2	0.0	5.0	-	-	2.3	5.0
3	X = 6.7	10	1.1	7.0	3.0	75.0	-	0.5	2.2	4.6
4	X = 8.9	36	1.1	10.0	30.4	75.0	0.5	0.1	1.7	5.2

Table 2. The formula of composition, the content of alloying elements, and the quantity of austenite in the MA alloys of series *B* after the ball milling and subsequent anneals.

No.	Formula of Composition	Ball Milling for 10 h			Annealing after BM at 700 °C (*a*) and 800 °C (*b*)					
	Fe*Y*Cr + 20Mn_2N (*Y* in wt%)	Quantity of Austenite γ, vol.%	Content of Nitrogen in Austenite γC_N, wt%	Contents of Cr and Mn in the α Phase αC_{Mn+Cr}, wt%	Quantity of Austenite γ, vol.%		Content of Nitrogen in Austenite γC_N, wt%		Contents of Cr and Mn in the α Phase αC_{Mn+Cr}, wt%	
					a	*b*	*a*	*b*	*a*	*b*
1	Y = 0.0	90	1.6	11.0	95	100	-	0.3	-	-
2	Y = 14.2	95	2.6	-	100	100	-	-	-	-
No.	Fe*Y*Cr + 10Mn_2N (*Y* in wt%)	Quantity of Austenite γ, vol.%	Content of Nitrogen in Austenite γC_N, wt%	Contents of Cr and Mn in the α Phase αC_{Mn+Cr}, wt%	Quantity of Austenite γ, vol.%		Content of Nitrogen in Austenite γC_N, wt%		Contents of Cr and Mn in the α Phase αC_{Mn+Cr}, wt%	
					a	*b*	*a*	*b*	*a*	*b*
3	Y = 0.0	10	1.1	4.0	28	10	0.4	0.4	2.0	4.0
4	Y = 4.7	24	1.0	8.7	24	80	0.1	0.1	2.5	5.5
5	Y = 8.6	36	0.7	11.5	31	100	0.3	0.3	4.6	-

Element formulas of compositions were selected based on the data from phase diagrams for the states of equilibrium of stainless austenitic steels, including those doped with nitrogen [21,22]. The binary alloys were smelted, homogenized, and then filed. The nitrides were synthesized using the technology of self-propagating high-temperature synthesis [23]. According to the data of X-ray diffraction (XRD) analysis, the chromium nitrides have presented by themselves the mixture of composition 80% CrN + 20% Cr_2N [18,19], and the manganese nitrides have been represented by 80% Mn_2N in mixture with the nitrides that have had an enhanced content of magnesium. According to the results of chemical analysis, the synthesized magnesium nitrides have contained 9.1 wt% (28.3 at.%) nitrogen [20]. Mechanical alloying was performed with the employment of a planetary ball mill «Pulverisette-7». The speed of rotation of the ball mill platform was 800 rpm. The vessel and balls were made of high-strength ball-bearing steel containing 1.5 wt% Cr and 1.0 wt% C, iron in balance. This steel is the most stable compared to materials such as tungsten carbide and stainless steel [24]. The weight ratio for the balls (the total quantity of 15 pieces, 10 mm in diameter each) and a powder sample was 6:1. After loading the balls and powder mixtures into the vessel, air was pumped out to 10^{-3} mm Hg and the vessel was filled with an inert gas-argon. The duration of milling time was varied from 5 to 20 h. The temperature at the external sides of the vessels did not exceed 70 °C. The average size of the particulates of the initial powders of the alloys was about 200 μm. After termination of ball milling, part of the powders were annealed in vacuum at temperatures of 700 °C (*a*) and 800 °C (*b*) for 1 h, to specify—in a vacuum of 10^{-5} mm Hg. Possible contamination of the test samples with wear products was controlled by weighing the mass of vessels, balls, and powder before and after mechanical alloying. The difference in the masses of powders did not exceed 0.3–0.5 wt%. The method of estimating powder contamination by weighing the mass of the vessels, balls, and powder is not absolute, i.e., it is not free of «errors». However, in this work, (i) taking into account the same conditions of exposure of different mixtures to mechanical processing and (ii) assuming the same systematic errors (in terms of the degree of contamination), the differences in the results of mechanical alloying of mixtures can be considered reliable. According to the results of chemical analysis, in all the cases of ball milling, the oxygen concentration in the MA samples did not exceed 0.8 wt%.

2.2. Mössbauer and TEM Analysis of MA Samples

The Mössbauer measurements of the product results of the MA and subsequent annealing of the powders were carried out at room temperature on the spectrometer unit MS-1101 in regime of constant accelerations with the source $^{57}Co(Rh)$. The spectra were calibrated using the absorbent material of α-Fe at a room temperature. Both the structure and the phase composition of an MA powder mixture were studied with the help of a transmission electron microscope JEM-200CX, with analyzing XRD patterns and dark-field images.

The Mössbauer spectra (transmission of gamma-quanta *A*, % as function of Doppler velocity *V*, mm/s) of MA samples have revealed in themselves a multicomponent structure typical of the solid solutions in both α-ferrite and γ-austenite phase states. Taking into account the complicated character of Mössbauer spectra, we used for their calculation the application-oriented soft-ware package MS Tools (borrowed from) [25]. Figures 1–5 show the experimental Mössbauer spectra and the results of their calculation. In particular, the program DISTRI was employed, which is usually used for—in the case of—locally inhomogeneous systems with a multicomponent structure and poor spectrum resolution. With its help one can improve the quality of a resolution of such spectra via restoring the distributions $p(V)$ and $p(H)$, which present by themselves the probabilities of resonance absorption represented on the scale of the Doppler velocities *V* and effective field *H*. Further, based on the (i) type of distributions $p(V)$ and $p(H)$, (ii) analysis of a priori data, and (iii) selection of the model for representing Mössbauer spectra, we made use of the soft-ware program SPECTR. For the calculating the parameters of the hyper-fine structure and partial contribution of the spectra components, we employed a standard procedure of approximating integral spectra by a superposition of the components with a Lorentzian shape of the lines.

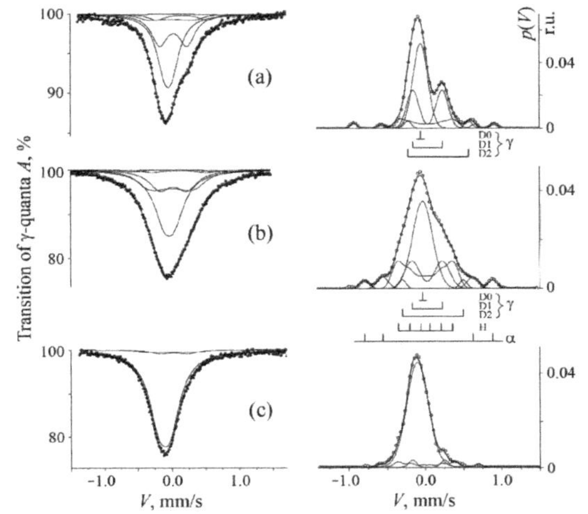

Figure 1. Mössbauer spectra and the distribution $p(V)$ for MA samples of Series B. Treatment, composition: (**a**) ball milling (BM), composition 2 (Fe–14.2Cr + 20% Mn_2N); (**b**) BM, composition 1 (Fe + 20% Mn_2N); (**c**) BM, annealing at 800 °C, 1 h, composition 1 (Fe + 20% Mn_2N). In the distributions $p(V)$ a "deciphering" of the model of nitrogen-containing austenite is given.

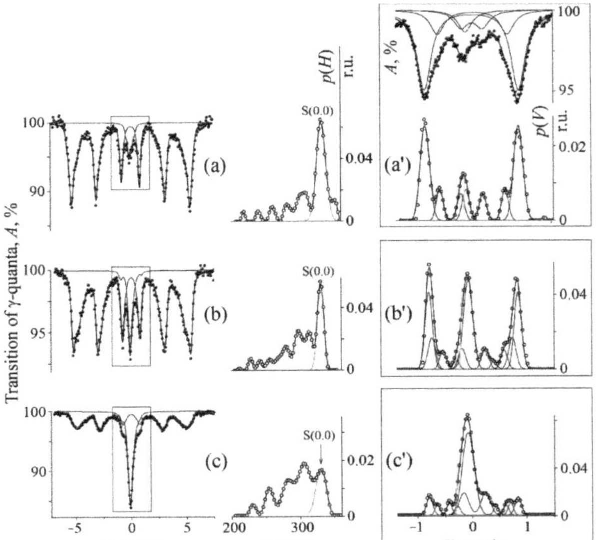

Figure 2. Mössbauer data of the MA samples after BM of Series A: (**a**–**c**) the Mössbauer spectra and distributions of the $p(H)$ of the sextet for the α phase; (**a′**) the Mössbauer spectra and distribution of the $p(V)$ of the center of the spectrum; (**b′**,**c′**) distributions of the $p(V)$ of the center of the spectra. Treatment, composition: (**a**,**a′**) BM, composition 1 (Fe + 20% CrN); (**b**,**b′**) BM, composition 3 (Fe–6.7Mn + 20% CrN); (**c**,**c′**) BM, annealing at 800 °C, 1 h, composition 4 (Fe–8.9Mn + 20% CrN).

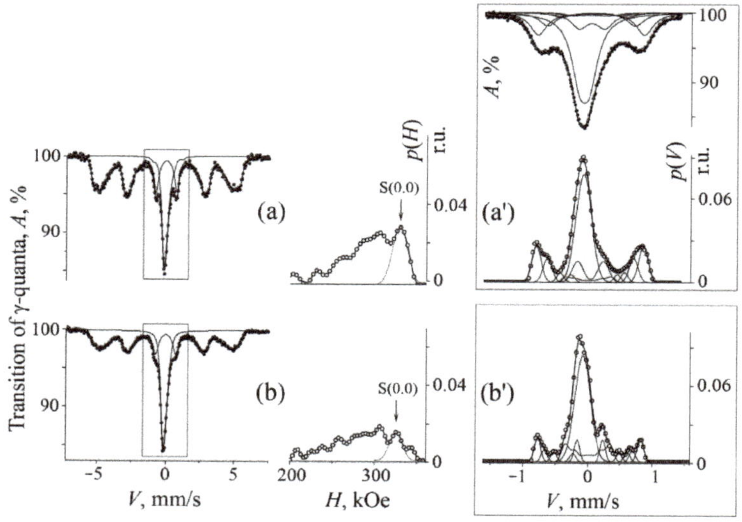

Figure 3. Mössbauer data of the MA samples after BM of Series *B*: (**a**,**b**) the Mössbauer spectra and distributions of the *p*(*H*) of the sextet for the α phase; (**a′**) the Mössbauer spectra and distribution of the *p*(*V*) of the center of the spectrum; (**b′**) distribution of the *p*(*V*) of the center of the spectrum. Treatment, composition: (**a**,**a′**) BM, composition 4 (Fe–4.7Cr + 10% Mn_2N); (**b**,**b′**) BM, composition 5 (Fe–8.6Cr + 10% Mn_2N).

For verification of the results of the mechanical alloying and modeling of the spectra of the austenite of MA alloys, special experiments were performed on the samples of the compositions 1 and 2 of the *B* series: Fe–14.2Cr + $20Mn_2N$ and Fe + $20Mn_2N$, see Table 2. Element formulas of the special compositions in the case of mechanical alloying permitted us to compare the obtained spectra with those well-known of the alloys produced by means of traditional metallurgy [26–36]. In the composition 1 from the series *B*, in accordance with the phase diagram borrowed from [21,22], in the case of MA one can expect the realization of the spectrum for paramagnetic nitrogen austenite, whereas in the composition 2 from the series *B*—one has the right to expect the realization of the spectrum of/from γ phase with a hyper-fine magnetic structure, the formation of which being the result of an anti-ferromagnetic (AFM) ordering in the Fe–Mn alloy with large (≥20%) content of manganese [29,32,33]. The results of calculation of the *p*(*V*) and spectra for the MA alloys of compositions 1 and 2 of (a) *B* series are shown in Figure 1a,b. The distribution *p*(*V*) reveals by itself a hyper-fine structure of the spectra of/from nitrided austenite, with that the spectra restored with the help of the data over *p*(*V*) describe the experiment well. Indeed, the spectra of/from the MA alloy of composition 1 of *B* series, judging by their shape and hyper-fine parameters agree well with the well-known data on the spectrum of/from nitrided paramagnetic austenite [16,30,31] (see Figure 1a and Table 3). In the case of samples of composition 2 from the series *B*, the measurements at room temperature show for the spectrum of/from the alloy the realization of broadening of the central singlet, which is characteristic of the AFM ordering in the FCC Fe–Mn austenite that contains manganese in quantity equal or greater (≥) than 20 wt% [29,32,34], Figure 1b and Table 1. To confirm the AFM origin of the broadening in the spectra of/from the MA alloy of composition 2 of series *B* there serves the emergence of the narrowing of the central singlet, which is due to the (i) partial decomposition of FCC solid solution and the (ii) exit of the manganese and nitrogen atoms from the matrix into nitrides, Figure 1c. When approximating the distribution *p*(*V*) by Gaussian forms, the model of the spectrum can be represented by the superposition of the components—$D(0) + D(1) + D(2) + S_{AFM}$ with the parameters given in Table 3. The selection of the Gaussian forms for the approximation of the lines and the very modeling itself are based on a priori info about the spectrum of the stainless steel [34,35] and nitrogen austenite [16,30,31]. The doublet $D(0)$ corresponds to the atoms of resonance-exhibiting iron without admixture of nitrogen, which is located in the octahedral interstitial positions—in the crystal lattice—of the first coordination shell (CS), and—by the magnitudes of the quadrupole shift and width of Gaussian forms—is close to the values of distribution (of probabilities of observation) of the electric field gradient and isomer shift for stainless steel [34,35]. The doublets $D(1)$ and $D(2)$ are of hyperfine parameters (namely, isomer shift I_s and quadrupole shift Qs) in value close to the parameters of the doublets from intrusion of one and two (so-called dumbbell configurations) atoms of nitrogen into the octahedral interstitial positions in the crystal lattice of FCC iron [30,31]. The sextet S_{AFM} with an hyperfine magnetic structure corresponds to a part of the formed γ phase with a large ($\geq$20%) content of manganese [32,33]. The spectra and distributions $p(V)$ are presented in Figure 1, and the parameters, in Table 3. Thus, special experiments with the samples of compositions 1 and 2 in the series B testify to the attainment of a desired mechanical alloying and to the possibility of modeling of spectra of MA alloys on the basis of models of Mössbauer spectra of nitrided austenite produced by traditional metallurgical methods [29–34].

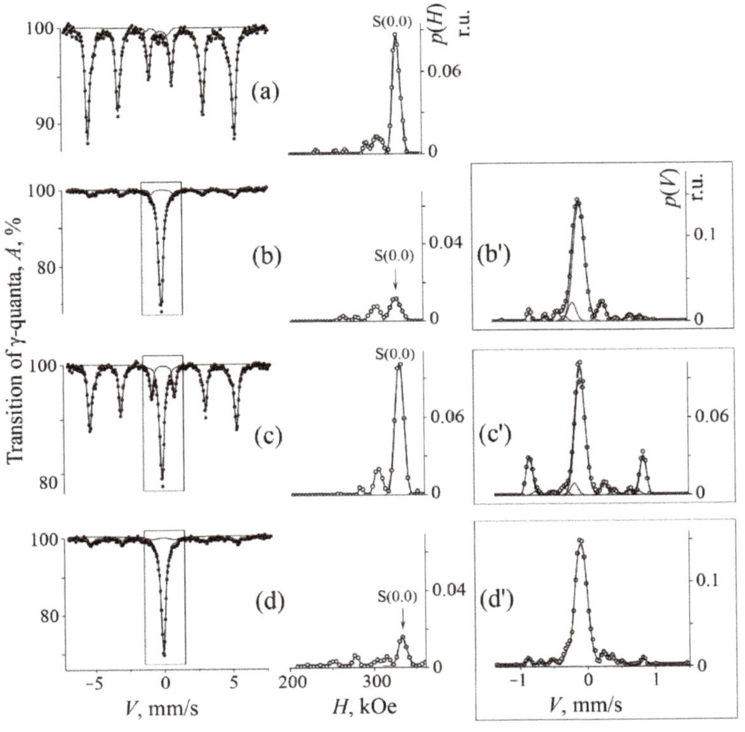

Figure 4. Mössbauer data of the MA samples after BM and annealing of Series *A*: (**a–d**) distributions of the $p(H)$ of the sextet for the α phase; (**b′–d′**) distributions of the $p(V)$ of the center of the spectra. Treatment, composition: (**a**) BM, annealing at 700 °C, 1 h, composition 3 (Fe–6.7Mn + 20% CrN); (**b,b′**) BM, annealing at 800 °C, 1 h, composition 3 (Fe–6.7Mn + 20% CrN); (**c,c′**) BM, annealing at 800 °C, 1 h, composition 4 (Fe–8.9Mn + 20% CrN); (**d,d′**) BM, annealing at 800 °C, 1 h, composition 4 (Fe–8.9Mn + 20% CrN).

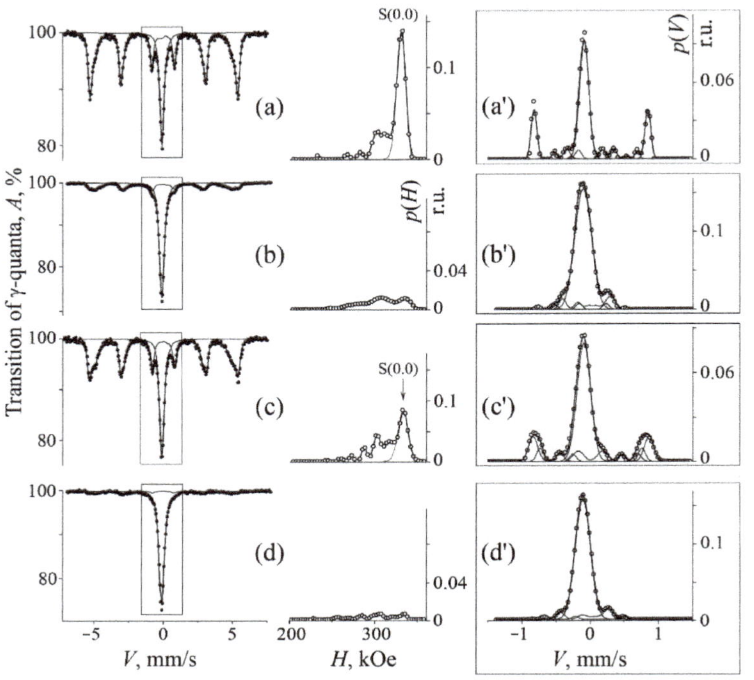

Figure 5. Mössbauer data of the MA samples after BM and annealing of Series *B*: (**a–d**) distributions of the $p(H)$ of the sextet for the α phase; (**a′–d′**) distributions of the $p(V)$ of the center of the spectra. Treatment, composition: (**a,a′**) BM, annealing at 700 °C, 1 h, composition 4 (Fe–4.7Cr + 10% Mn_2N); (**b,b′**) BM, annealing at 800 °C, 1 h, composition 4 (Fe–4.7Cr + 10% Mn_2N); (**c,c′**) BM, annealing at 700 °C, 1 h, composition 5 (Fe–8.6Cr + 10% Mn_2N); (**d,d′**) BM, annealing at 800 °C, 1 h, composition 5 (Fe–8.6Cr + 10% Mn_2N).

Table 3. Hyper-fine parameters of simulated spectra of the MA nitrogen-containing austenite, stainless austenite Fe–22Mn–18Cr–0.8N steel, and nitrogen-containing γ iron.

Formula of Composition	Doublet $D(0)$		Doublet $D(1)$		Doublet $D(2)$		Sextet S_{AFM}			Comments and Refs
	I_S, mm/s ±0.01	Q_S, mm/s ±0.01	I_S, mm/s ±0.01	Q_S, mm/s ±0.01	I_S, mm/s ±0.01	Q_S, mm/s ±0.01	I_S, mm/s ±0.02	Q_S, mm/s ±0.02	H, kOe ±2	
Fe + 20% CrN	−0.07	0.00	0.03	0.19	0.18	0.41	-	-	-	Series *A*, composition 1
Fe-14.2Cr + 20% Mn_2N	−0.05	0.14	0.03	0.20	0.16	0.36	–	-	-	Series *B*, composition 2
Fe + 20% Mn_2N	−0.05	0.06	0.03	0.20	0.16	0.37	0.00	0.00	22	Series *B*, composition 1
Fe-4.7Cr + 10% Mn_2N	−0.11	0.00	0.03	0.20	-	-	−0.01	−0.04	24	Series *B*, composition 4 annealing
Fe-9.0N	−0.01	0.00	0.08	0.20	0.20	0.36	-	-	-	[30]
Fe-8.5N	−0.02	0.00	0.06	0.20	0.31	0.33	-	-	-	[31]

Spectra of/from powders of the initial ferromagnetic alloys Fe–*X*Mn and Fe–*Y*Cr, as well as of the α phase in the body of MA samples, present by themselves the sextets with broadened lines which are a superposition of the subspectra $S(n1, n2)$ correspondent to the non-equivalent surroundings of atoms of resonance-exhibiting iron by the impurities of chromium and manganese in the positions (lattice sites) of substitution in the solid solutions with BCC crystal lattice [26–28]. In result of the mechanical alloying with the formation of solid solutions, along with the sextet from the α phase, in the center of the spectra there appears a broadened asymmetrical singlet from austenite. In a first approximation (without taking into account the difference in the values of Debye–Waller factor of phase components), the numerical ratio of the areas under the integral sextet and central singlet corresponds—and is equal in volume percentages—to the quantitative ratio of the volumes of the α and γ phases in the MA alloy.

Quantitative assessment of the nitrogen content C_N (in at.%) in austenite was carried out, similarly to that performed in [15,16,30] under the assumption of repulsive distribution (i.e., mutual repulsion) of nitrogen atoms in a solid solution, by the contribution of the atomic configuration represented by the $D(1)$ doublet (Figure 1) from iron atoms with one nitrogen atom in the nearest octahedral interstitial positions. The value of the integral intensity from the doublet $D(1)$ is related to C_N by the relationship [30]:

$$S_{D(1)} = 6p(1 - p),$$

where $p = C_N(1 - C_N)$ is the fraction of the octahedral interstitial sites in austenite occupied by nitrogen atoms. In order to account for the self-absorption effect, the intensity of $S_{D(1)}$ was extrapolated to the zero thickness of the absorber. The change in the integral intensity of the doublet $D(1)$ can be explained by the self-transfer of nitrogen from nitrides to the solid solution of γ phase in the course of deformation-induced dissolution and vice versa during the decomposition of solid solution under conditions of nitrogen binding with chromium or manganese into next-emerging secondary nitrides [15,16].

Calculations of the spectra of/from the ferrite α matrix in the MA alloys and, in particular, of the quantity of impurity of substitution in this matrix were performed in the assumptions that (i) the contributions of chromium and manganese to the change in the isomer shift I_s and effective field H at Fe atoms was additive in nature and (ii) the (binomial) distribution of impurities (namely, Cr and Mn) in the solid solution was of chaotic character [26–28,36]. Estimation of the total content of the chromium and manganese in the solid solution of ferrite matrix of the MA alloy, αC_{Cr+Mn}, was performed via establishment of a partial contribution to the spectrum and distribution $p(H)$ from the impurity-free atomic surroundings of resonance-exhibiting iron—i.e., from the sextet $S(n1, n2)$ with $n1 = n2 = 0$, where $n1$ and $n2$ are the quantities (number) of atoms of chromium and manganese in the first and second CS, respectively. The sextet $S(0,0)$ in the diluted Fe–Mn and Fe–Cr solid solutions is easily separated in the form of the peak in the distribution $p(H)$ with magnitudes of fields in the vicinity of 330 kOe, see Figure 2a. Nitrogen in the ferromagnetic component of the MA alloy weakly affects (i) the hyper-fine parameters of the spectrum of α solid solution, as well as (ii) the value of the parameter of the mean effective field $\langle H \rangle$ [17]. Therefore, we present no quantitative estimations of the nitrogen content in the α solid solution of the MA alloy.

3. Mechanical Alloying in the Course of the Ball Milling and Thermal Annealing

3.1. Mössbauer Analysis of Mechanical Alloying in the Cases of A and B Series

In the case of samples of the series *A*: Fe–XMn + 20CrN (Cr_2N), the results for MA mixtures in the compositions with X = 0, 6.7, 8.9 wt% Mn are presented in Figure 2 and in Table 1. It is seen that an increase of the concentration of manganese in the initial matrix from 0 to 8.9 wt% leads to the growth of intensity of the central singlet of paramagnetic austenite from 6 to 36 vol.%. In its turn, in the singlet of/from the austenite, the doublets $D(1)$ and $D(2)$ are well pronounced, which testifies that (i) nitrogen «hits» into the octahedral interstitial positions in the crystal lattice of FCC phase and (ii) nitrogen-containing austenite is formed. In the spectrum for the α phase—being a ferrite component of the MA alloy, changes also are observed, namely, there takes place a pronounced lowering in the magnitude of the mean effective magnetic field $\langle H \rangle$ of the integral sextet at the expense of lowering of a partial contribution to the spectrum and $p(H)$ of the superpositional sextet $S(n1,n2)$ with $n1 = n2 = 0$, correspondent to the impurity-free surroundings of iron by the atoms of chromium and manganese [17,18,26,36]. The estimation of the content of impurity of substitution in the MA α solid solution testifies to the increase in the concentration of the chemical elements of substitution in the matrix of α phase, see the results of BM in Table 1. Thus, based on the spectra results of the change—in result of ball milling (BM)—in the ferrite and austenite components in the samples of series *A*, the chemical elements from the initial nitrides CrN partially pass to the body of the α and γ solid solutions.

In the case of samples of the series *B*: Fe–*Y*Cr + $10Mn_2N$, the results for the MA mixtures are presented in Figure 3 and in Table 2. The spectra for the MA results have qualitatively the same "appearance" as in the case of series *A*. In the sextet that describes a ferromagnetic α solid solution, one can observe the decrease in the magnitude of <*H*> as a consequence of the enrichment of metallic matrix by manganese, i.e., the dissolution of nitrides has taken place and passing of manganese atoms from the nitrides to the α matrix of the MA alloy has occurred in parallel. However, a noticeable quantitative distinction of the experimental results exists relative to those for the series *A*.

This, first of all, concerns the increase (i) in the volume of austenite, as well as (ii) of the degree of alloying of the α phase by manganese. In the case of the samples of the series *B* with the composition formulas 3 and 4 (see Table 2), the ball milling preceding sample compaction brings about the quantity of austenite from 10 to 24 vol.%, whereas in the case of the samples of the series *A* with the composition formulas from 1 to 3 (see Table 1) the processing mentioned entails the increase in the quantity of austenite component from 4 to 7 vol.%. At the same time, the content of alloying elements of substitution in the ferrite matrix, αC_{Mn+Cr}, increases in a more pronounced extent in comparison with that one which is characteristic of the case of series *A*. Besides, in the central singlet of/from austenite the component related with the AFM ordering is distinguished to stand obviously apart, see Figure 3a,c. This testifies to the formation in the γ-phase of the regions enriched in manganese, i.e., having the manganese content greater than that is characteristic of the average for the composition. The values of the parameters for the sextets "typical" of the AFM ordering are presented in Table 3. The content of nitrogen in austenite for the series *A* and *B* amounts from 0.7 wt (2.6 at.) to 3.7 wt (13.1 at.) %.

3.2. Mössbauer Analysis of MA Samples after Their Annealing

In the case of samples of the series *A*, the result of an annealing of the MA mixtures at 700 °C consists in the reverse evolution of the spectra with considerable lowering of the intensity of the central singlet of/from austenite, Figure 4a,c. The content of substitution impurity in the α solid solution approaches the initial—or a lower one than that is characteristic of the initial binary Fe–*X*Mn alloy, Table 2 *a*. In particular, in the composition Fe–8.9Mn + 20CrN, after the MA and annealing at 700 °C, for 1 h, the total content of the manganese and chromium in the α phase does not exceed (i) 1.7 wt% at 8.9 wt% Mn in the initial binary alloy and (ii) 10.0 wt% in the MA alloy after BM. The quantity 36 vol.% of austenite remains virtually the same (i.e., unaltered), but the content of nitrogen exhibits a lowering from 1.1 wt (4.2 at.) % to 0.5 wt (1.8 at.) %. An annealing at a higher temperature, at 800 °C, for 1 h, facilitates the occurrence of the $\alpha \rightarrow \gamma$ transition in the samples of compositions 3 and 4 with $X = 6.7$ and 8.9 and makes lower the degree of decomposition of the solid solutions formed in the course of MA. In samples of all the compositions of series *A*, an annealing at a higher temperature preserves the values of <*H*> and of the intensity of *S*(0,0) component of the sextet of/from the α phase from alteration, i.e., the annealing stabilizes the chemical elements of the substitution in the α matrix. The intensity of *D*(1) decreases in a lesser extent, i.e., nitrogen is partly preserved in the solid solution of paramagnetic austenite, see Figure 4b,d and Table 1 *b*.

In the case of series *B*, the annealing of the MA samples differs, first of all, by increasing the quantity of austenite in the cases of all compositions, see Figure 5 and Table 2. After annealing at 800 °C of the samples of composition 5 with $Y = 8.6$ wt%, up to 100% of austenite is formed. In the austenite(-stipulated) singlet of/from the MA alloy one can observe the preservation of the component related to the AFM ordering, which is described by the sextet with the combined magnetic and quadrupole splitting, see Table 3. The same but in a better way as it takes place in the series *A*, annealing at 800 °C stabilizes the elements of mechanical alloying in the ferrite matrix; the comparison of the results of annealing of samples of the compositions in the series *A* and *B* is shown in Table 2.

3.3. TEM Data on the Results of the MA and Subsequent Annealing

According to the TEM data on the MA alloyed powder mixtures *A* and *B*, the same as in the case of mixtures based on the Fe–Ni matrix [18], one observes there the formation of a submicrocrystalline structure (SMC). In Figure 6 the dark-field image and SAED pattern of the powder mixture Fe–8.6Cr + 15Mn_2N are presented. The SAED pattern (Figure 6b) contains ring (circular-wise disposed) reflections from greatly misoriented grains of austenite. According to [37], reflections from the nitrides CrN and Cr_2N are located near the reflections from austenite. In the dark-field image taken in the reflection «austenite + nitrides» one can observe the images of the nitrides and fragments of austenite. The size of the fragments of matrix (grains and subgrains) amounts to 50–80 nm. Also visible are more disperse (finer) precipitates of size about 2 nm, which correspond to the secondary nitrides of the types CrN and Cr_2N.

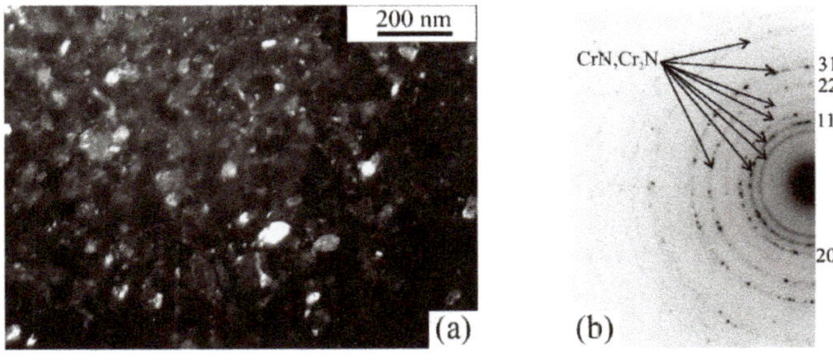

Figure 6. TEM structure of the MA powder mixture Fe–8.6Cr + 15Mn2N: (**a**) dark-field image taken in the combined reflection ($(111)_\gamma$ + CrN, Cr_2N); (**b**) SAED pattern.

After termination of the annealing at 800 °C for 1 h, an SMC structure of the powder mixture remained unaltered, see Figure 7. The phase composition of the powder mixture Fe–8.6Cr + 15Mn_2N presents by itself—after the milling and annealing—the austenite and chromium nitrides (CrN and Cr_2N). Annealing of the MA samples leads both to an increase in the size of chromium nitrides, which have been formed in the course of mechanical alloying, and to their precipitation at the dissolution of supersaturated γ solid solution. In the SAED pattern shown in Figure 7a, one can see two first reflection rings that correspond to nitrides of chromium (CrN and Cr_2N). In the dark-field image taken in the reflection «CrN + Cr_2N», (Figure 7b), it is these nitrides that shine.

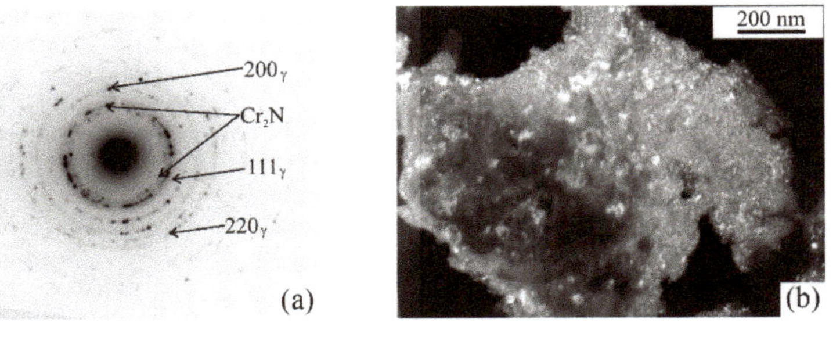

Figure 7. TEM structure of the MA powder mixture Fe–8.6Cr + 15Mn_2N after annealing at 800 °C for 1 h: (**a**) SAED pattern; (**b**) dark-field image taken in the combined reflection (CrN, Cr_2N).

4. Discussion

Using as metallic matrix precisely the Fe–Mn and Fe–Cr alloys instead of the pure metals Fe, Mn, and Cr with admixture of nitrides, has its own peculiar features that affect the (i) mechanism of structure-phase transitions, (ii) kinetics, and (iii) properties of the products of MA. This influence is stipulated by the physicochemical properties of the components under alloying and, in particular, by the phase diagram of the alloys and nitrides, as well as (by) the difference in the chemical activity and diffusion mobility of the elements of alloying (Cr, Mn, and N), in the course of the MA and subsequent annealing.

As it follows from the results of Mössbauer investigations, in the samples of series *A* and *B* one can observe—under BM—the smelting of the initial binary Fe–Mn and Fe–Cr alloys, on the one hand, and the nitrides CrN and Mn_2N, on the other. Experiments on the mechanical alloying in ball mills—used in the capacity of metallic matrix the alloys, —are of common regularities, with SPD-induced (by severe plastic deformation) structure-phase transformations upon the friction- and HPT-providing treatments executed on the nitrided surface of the iron alloys and stainless steels [15–17]. In the course of severe deformation, in MA samples we encounter the formation of solid solutions with the element and phase composition similar to that typical of Fe–Cr–Mn–N alloys produced in a traditional metallurgical way, see data of Tables 1 and 2. The ferromagnetic component having BCC crystal lattice also exhibits a change in its composition by revealing the altering of the quantity of the manganese and chromium in the α solution. In result of an annealing at 700 and 800 °C of the MA samples, both in the series *A* and in the series *B*, there occurs the $\alpha \rightarrow \gamma$ phase transition accompanied by (i) the decomposition of the formed supersaturated solid solutions and (ii) by the exit of nitrogen and of the alloying elements manganese and chromium from the metallic matrices into secondary nitrides, see Tables 1 and 2 *a*, *b*. Moreover, at BM and annealing, in the processes of decomposition of MA solid solutions there participate the chemical elements of an alloying (by chromium and manganese) of the initial binary alloys. After annealing, the content of the chromium and manganese in the α matrix of MA alloy is noticeably decreased in comparison with the content of these elements in the alloys of initial composition. For instance, in the case of samples of composition 4 (Fe–8.9Mn + 20CrN) of *A* series, the total content of substitution elements in the α phase, αC_{Mn+Cr}, decreases to 1.7 wt%, whereas the concentration of nitrogen in austenite does not exceed 0.5 wt (2 at.)%. In the case of series *B*, at 700 °C a qualitatively similar pattern (picture) is observed, see data for the samples of compositions 4 and 5 in Table 2 and in Figure 5a,c. We can assume that active exit of substitution elements from the α matrix is connected not only with the decomposition of supersaturated solid solutions and the formation of nitrides, but also with the diffusion exchange of the chromium and manganese atoms between the α and γ phases in accordance with the phase diagrams of these alloys. Moreover, the nanostructurization of metallic matrix in the course of BM contributes to the formation of segregations of the chromium, manganese, and nitrides on along the developed grain boundaries [38,39]. The structure of the nitrogen-containing MA austenite after the thermal annealing and cooling to room temperature experiences a better stabilization precisely after annealing at a higher temperature, namely, of 800 °C, see Tables 1 and 2 *a*, *b*. It can be seen from the Tables that the stabilization of nitrogen-containing austenite at the annealing of higher temperature is explained by the presence of chemical elements of mechanical alloying that have finally remained in the body of a volume of the solid solution.

The data of TEM have confirmed the validity of results of Mössbauer measurements concerning the formation of nitrogen-containing austenite together with the secondary fine nitrides (CrN and Cr_2N) in the course of MA of the powder mixtures Fe–Mn–*X*CrN and Fe–Cr–*Y*Mn_2N. At high-temperature anneals there occurs both the increase in the size of chromium nitrides that have been formed at mechanical alloying of powder mixtures and the precipitation of new fine ones during the decomposition of supersaturated γ solid solution. The preservation at high-temperature annealing of the submicro- and nano-sized austenite grains in MA alloys can be in part explained by the formation of the secondary nitrides CrN and Cr_2N along the developed grain boundaries, which evolve in parallel of the $\alpha \rightarrow \gamma$ phase transition, slows down the process of recrystallization-controlled growth of grains, see Figure 7a,b.

The alloys obtained by the MA and subsequent annealing at 700 and 800 °C, are metastable and at cooling to room temperature they can pass by the mechanism of shear transformation just to the α phase. The small size of austenite grains [40], short-range-order atomic concentrational separation [41], and precipitation of disperse nitrides coherently matched with the matrix [42], all this play, in the case under consideration, a stabilizing role in relation to the occurrence of the $\gamma \rightarrow \alpha$ transition. An additional stabilization of austenite can be connected with the presence of the concentrational inhomogeneity over a material. The stabilizing role of concentrational inhomogeneities was exemplified on the austenitic Fe–Ni alloys [43]. The formation of the concentrational inhomogeneities over manganese in the austenitic MA samples of series *B*, and more specifically, of the regions of the γ phase with an increased content of magnesium, appears both after ball milling and after annealing right in the form of the component related with the AFM ordering, see Figures 1c and 5b. In the spectrum for the alloy Fe–4.7Cr + $10Mn_2N$ annealed at 800 °C, for 1 h, these regions of a structure are described by the sextet with a combined magnetic and quadrupole splitting, see Figure 5b,d and Table 1.

From the results of the BM and anneals of samples from the series *A* and *B*, it follows that the series *B* turns out to be more efficient from the point of view of the volume of a stable austenite being under formation at the lesser (i) quantity *of* alloying elements (*of* chromium, manganese, and nitrogen) and (ii) amount of the primary nitrides in the samples of these compositions, see Table 2. A more efficient formation of the nitrogen-containing austenite and its stabilization in the samples of series *B* can be explained by several reasons.

First, (it can be explained) (i) by a more efficient dissolution of Mn_2N nitrides in a metallic matrix and (ii) by the formation of the Mn-rich austenite-forming regions in a metallic matrix of MA alloy with the concentration of manganese ~20 at. %. Manganese-rich regions of austenite can be formed by the actual smelting of nitrides both with austenite and with a paternal α phase. A more active alloying of the initial matrix in the samples of series *B* is well illustrated by data from Tables 1 and 2. The manganese-rich regions of the α phase in this very case possess an enhanced stimulus to/for the $\alpha \rightarrow \gamma$ transition in accordance with the phase diagram (in) [21,22]. Accommodation stresses between (i) the γ phase enriched in manganese during its formation in the course of MA treatment and (ii) the rest matrix, stabilize austenite in relation to the $\gamma \rightarrow \alpha$ transition [42,43].

Second, the high efficiency of the formation of austenite in the samples of series *B* can be explained by a low diffusion mobility of the atoms of manganese (having a larger atomic radius in comparison with that of chromium), which (i) makes weaker the process of dynamic aging (at BM) concurrent with the process of dissolution of nitrides and (ii) facilitates the preservation of manganese to remain in the solid solution. The (in-case) presence of the dynamic aging concurrent with the dissolution is obvious from the slight chromium doping of the initial Fe–XMn matrix, see the result of BM in Table 1 that presents data for the case of series *A*. The accelerated concurrent process in the samples of series *A* presumably is explained by the enhanced chemical and diffusion activity of chromium. It is noticeable that it is in the case of series *B* there more actively occur both dissolution of Mn_2N nitrides and alloying of the α phase by manganese, see Table 2. The formation of the regions enriched in manganese is most distinctly revealed in conditions of the BM at relatively moderate temperatures. Synergetic nature of deformation-induced processes, which manifests itself in the decomposition of supersaturated solid solutions in the course of concurrent mechanical alloying, was demonstrated in the works on the short-range order investigations into the Fe–Ni and Fe–Cr alloys [44,45]. The influence of the diffusion mobility of the alloying elements of substitution on this process was shown in the works on the dynamic aging in the Fe–Ni–*Me* alloys where *Me* = Ti, Al, Si, Zr [44,46,47].

The third reason for the high efficiency of the formation of austenite in the case of series *B* consists in the lower thermal stability of Mn_2N nitrides in comparison with that of the nitrides CrN (Cr_2N) in conditions of the local heating-up of the mixture at the BM and high-temperature anneals. According to the data from [48], the nitride Mn_2N undergoes decomposition at 900 °C, while the nitride CrN decomposes at sufficiently higher temperatures (>1000 °C). The effect of the thermal stability and chemical activity of the constituents of would-be MA products on the kinetics of the dynamic aging and formation of secondary phases was demonstrated in the example of deformation-induced dissolution of nitrides, as well as intermetallics in [44,46,47].

5. Conclusions

Using the methods of Mössbauer spectroscopy and electron microscopy, we have investigated the structure-phase transitions in the course of solid-phase mechanical alloying in a ball mill and at subsequent annealing of the mixtures of the Fe–Cr and Fe–Mn alloys, on the one hand, and the nitrides CrN (Cr_2N) and Mn_2N, on the other. The samples from the series *A*, of compositions Fe–*X*Mn + 20CrN (Cr_2N), and the samples from the series *B*, of compositions Fe–*Y*Cr + 10, 20Mn_2N, —were the objects of investigation. It is shown that the ball milling of the mixtures from the series *A* and *B* leads to the smelting of the components of mixtures and to the formation of the Fe–Cr–Mn–N solid solutions with FCC and BCC crystal lattices. Increasing the concentration of the elements Cr and Mn in the initial BCC alloys, as well as the proportion amount of nitrides in the studied compositions in conditions of the ball milling and subsequent annealing, is accompanied by the increase in the volume of forming austenite with nitrogen concentration of 3.7 wt (13.1 at.)%. A result of the mechanical alloying of samples from the series *B* is the formation of the FCC phase, part of which is of AFM ordering development with the content of manganese equal or greater ($\geq$) than 20 wt%.

The process of the ball milling and subsequent isothermal annealing of the MA mixtures correspondent to the series *A* and *B* is accompanied by the $\alpha \rightarrow \gamma$ phase transitions, formation of nitrogen-supersaturated solid solutions, and their decomposition with the formation of secondary nitrides. Austenite formed in the course of ball milling and subsequent annealing at 700 and 800 °C, for 1, has submicro- and nano-sized grains, which along with the coherent disperse (fine) secondary nitrides and alloying elements are preserved in the matrix, which is a reason for the stabilization of austenite. It has been established experimentally that annealing of MA samples at 700 °C leads to a sharp decrease in the amount of the alloying elements of manganese and chromium in the α phase, which along with the decomposition and formation of secondary nitrides can be a consequence of diffusion of chromium and manganese from the volume of the α into γ phase. It is shown that the *B* series—in its samples with the primary nitrides Mn_2N—is both at ball milling and at subsequent annealing more efficient from the viewpoint of forming the greater volume of MA austenite and its saturation with nitrogen at a lesser—than is characteristic of the case of the series *A*—content of the alloying chromium and manganese. The higher stability of MA austenite in the case of the series *B* is explained by the (i) formation of concentration-inhomogeneous regions of the matrix enriched in austenite-forming manganese, (ii) thermal stability of Mn_2N nitrides being lower in comparison with that of CrN (Cr_2N), and (iii) low diffusion mobility of manganese in the course of the ball milling and subsequent annealing.

Author Contributions: Conceptualization, V.S.; methodology, V.S., E.N., K.L., A.Z., N.K. and K.K.; validation, V.S., K.L., N.K. and K.K.; formal Analysis, V.S., A.Z., N.K., K.L. and K.K.; investigation, K.L., K.K. and E.N.; writing—original draft preparation, V.S. and K.L.; writing—review and editing, Y.U., K.L. and K.K.; project administration, V.S.; funding acquisition, K.L. and V.S. All authors have read and agreed to the published version of the manuscript.

Funding: The reported study was funded by the Russian Foundation for Basic Research (project no. 19-33-60006). The research was carried out within the state assignment of Ministry of Science and Higher Education of the Russian Federation (theme "structure" no. AAAA-A18-118020190116-6.

Acknowledgments: The authors are grateful to Gennady Dorofeev for the providing of nitride powders and their attestation.

Conflicts of Interest: The authors declare no conflict of interest.

References

1. Rashev, T.V. *High Nitrogen Steels. Metallurgy under Pressure*; Bulgarien Academy of Science: Sofia, Bulgaria, 1995.
2. Simmons, J.W. Overview: High-nitrogen alloying of stainless steels. *Mater. Sci. Eng. A* **1996**, *207*, 159–169. [CrossRef]
3. Yang, K.; Ren, Y. Nickel-free austenitic stainless steels for medical applications. *Sci. Technol. Adv. Mater.* **2010**, *11*, 014105. [CrossRef] [PubMed]
4. Mendez, M.; Mancha, H.; Mendoza, G.; Escalante, J.I.; Cisneros, M.M. Structure of a Fe-Cr-Mn-Mo-N alloy processed by mechanical alloying. *Metall. Mater. Trans. A* **2002**, *33*, 3273–3278. [CrossRef]
5. Fukutsuka, T.; Anzai, T.; Kaneda, M.; Matsuo, Y.; Sugie, Y.; Fukaura, K. Preparation of High Nitrogen Containing Stainless Steels by Mechanical Alloying Method and Their Localized Corrosion Behavior. *J. Soc. Mater. Sci. Jpn.* **2004**, *53*, 1175–1179. [CrossRef]
6. Cisneros, M.M.; López, H.F.; Mancha, H.; Rincón, E.; Vázquez, D.; Pérez, M.J.; de la Torre, S.D. Processing of nanostructured high nitrogen stainless steel by mechanical alloying. *Metall. Mater. Trans. A* **2005**, *36*, 1309–1316.
7. Guan, L.; Qu, X.; Wang, S. Preparation of stainless steel powder containing nitrogen by mechanical alloying technique. *J. Univ. Sci. Technol. Beijing* **2005**, *27*, 692–694.
8. Dawei, C.; Xuanhui, Q.; Ping, G.; Li, K. Preparation of nearly spherical nickel-free highnitrogen austenitic stainless steel powders by mechanical alloying. *Powder Metall. Technol.* **2008**, *26*, 265–268.
9. Amini, R.; Hadianfard, M.J.; Salahinejad, E.; Marasi, M.; Sritharan, T. Microstructural phase evaluation of high-nitrogen Fe–Cr–Mn alloy powders synthesized by the mechanical alloying process. *J. Mater. Sci.* **2009**, *44*, 136–148. [CrossRef]
10. Haghir, T.; Abbasi, M.H.; Golozar, M.A.; Panjepour, M. Investigation of α to γ transformation in the production of a nanostructured high-nitrogen austenitic stainless steel powder via mechanical alloying. *Mater. Sci. Eng. A* **2009**, *507*, 144–148. [CrossRef]
11. Salahinejad, E.; Amini, R.; Hadianfard, M.J. Contribution of nitrogen concentration to compressive elastic modulus of 18Cr–12Mn–XN austenitic stain less steels developed by powder metallurgy. *Mater. Des.* **2010**, *31*, 2241–2244. [CrossRef]
12. Salahinejad, E.; Amini, R.; Askari, B.E.; Hadianfard, M.J. Microstructural and hardness evolution of mechanically alloyed Fe–Cr–Mn–N powders. *J. Alloys Compd.* **2010**, *497*, 369–372. [CrossRef]
13. Salahinejad, E.; Amini, R.; Hadianfard, M.J. Structural evolution during mechanical alloying of stainless steels under nitrogen. *Powder Technol.* **2012**, *2015–2016*, 247–253. [CrossRef]
14. Tsuchiyama, T.; Uchida, H.; Kataoka, K.; Takaki, S. Fabrication of Fine-grained High Nitrogen Austenitic Steels through Mechanical Alloying Treatment. *ISIJ Int.* **2002**, *42*, 1438–1443. [CrossRef]
15. Shabashov, V.A.; Borisov, S.V.; Zamatovsky, A.E.; Vildanova, N.F.; Mukoseev, A.G.; Litvinov, A.V.; Shepatkovsky, O.P. Deformation-induced transformations in nitride layers formed in bcc iron. *Mater. Sci. Eng. A* **2007**, *452–453*, 575–583. [CrossRef]
16. Shabashov, V.A.; Korshunov, L.G.; Sagaradze, V.V.; Kataeva, N.V.; Zamatovsky, A.E.; Litvinov, A.V.; Lyashkov, K.A. Mössbauer analysis of deformation dissolution of the products of cellular decomposition in high-nitrogen chromium manganese austenite steel. *Philos. Mag.* **2014**, *94*, 668–682. [CrossRef]
17. Shabashov, V.A.; Kozlov, K.A.; Lyashkov, K.A.; Litvinov, A.V.; Dorofeev, G.A.; Titova, S.G. Solid-Phase Mechanical Alloying of BCC Iron Alloys by Nitrogen in Ball Mills. *Defect Diffus. Forum* **2012**, *330*, 25–37. [CrossRef]
18. Shabashov, V.A.; Kozlov, K.; Lyashkov, K.A.; Kataeva, N.V.; Litvinov, A.V.; Sagaradze, V.V.; Zamatovskii, A.E. Solid-state mechanical synthesis of austenitic Fe-Ni-Cr-N alloys. *Phys. Met. Metallograph.* **2014**, *115*, 392–402. [CrossRef]
19. Dorofeev, G.A.; Sapegina, I.V.; Lad'Yanov, V.I.; Pushkarev, B.E.; Pechina, E.A.; Prokhorov, D.V. Mechanical alloying and severe plastic deformation of nanocrystalline high-nitrogen stainless steels. *Phys. Met. Metallograph.* **2012**, *113*, 963–973. [CrossRef]
20. Dorofeev, G.A.; Lubnin, A.N.; Ulyanov, A.L.; Mukhgalin, V.V. Accelerated mechanosynthesis of high-nitrogen stainless steel: Mössbauer and X-ray diffraction studies. *Bull. Russ. Acad. Sci. Phys.* **2017**, *81*, 803–806. [CrossRef]
21. Gavriljuk, V.G.; Berns, H. *High Nitrogen Steel: Structure, Properties, Manufacture, Applications*; Springer: Berlin/Heidelberg, Germany, 1999.
22. Kirchner, G.; Uhrenins, B. Experimental study of the ferrite/austenite equilibrium in the Fe-Cr-Mn system and the optimization of thermodynamic parameters by means of a general mathematic method. *Acta Metall.* **1974**, *22*, 523–532. [CrossRef]
23. Dorofeev, G.A.; Karev, V.; Goncharov, O.; Kuzminykh, E.; Sapegina, I.; Lubnin, A.; Mokrushina, M.; Lad'Yanov, V. Aluminothermic Reduction Process Under Nitrogen Gas Pressure for Preparing High Nitrogen Austenitic Steels. *Met. Mater. Trans. B* **2019**, *50*, 632–640. [CrossRef]
24. Konygin, G.N.; Stevulova, N.; Dorofeev, G.A.; Elsukov, E.P. The effect of wear of grinding tools on the results of mechanical alloying of Fe and Si(C). *Chem. Sustain. Dev.* **2002**, *10*, 73–80.
25. Rusakov, V.S.; Kadyrzhanov, K.K. Mössbauer Spectroscopy of Locally Inhomogeneous Systems. *Hyperfine Interact.* **2005**, *164*, 87–97. [CrossRef]
26. Kang, S.G.; Onodera, H.; Jamamoto, H.; Watanabe, H. Mössbauer effect study of Fe-Mn alloys. *J. Phys. Soc. Japan* **1974**, *36*, 975–979. [CrossRef]
27. Vincze, I.; Campbell, I.A. Mössbauer measurements in iron based alloys with transition metals. *J. Phys. F Met. Phys.* **1973**, *3*, 647–663. [CrossRef]

28. van der Woude, F.; Sawatzky, G.A. Mössbauer effect in iron and dilute iron based alloys. *Phys. Rep.* **1974**, *12*, 335–374. [CrossRef]
29. Pepperhoff, W.; Acet, M. *Constitution and Magnetism of Iron and its Alloys*; Springer: Berlin/Heidelberg, Germany, 2001.
30. Oda, K.; Umezu, K.; Ino, H. Interaction and arrangement of nitrogen atoms in FCC γ-iron. *J. Phys. Condens. Matter* **1990**, *2*, 10147–10158. [CrossRef]
31. Ino, H.; Umezu, K.; Kajiwara, S.; Uehara, S. Interstitial solute atom configuration in Fe-N and Fe-C based austenite and relation to the abnormal tetragonality of fresh martensite. In Proceedings of the International Conference on Martensitic Transformations (ICOMAT-86), Nara, Japan, 26–30 August 1986; The Japan Institute of Metals: Sendai, Japan, 1986; pp. 313–318.
32. Endoh, Y.; Ishikawa, Y. Antiferromagnetism of γ–iron–manganes alloys. *J. Phys. Soc. Jpn.* **1973**, *30*, 1614–1627. [CrossRef]
33. Amigud, G.G.; Bogachev, I.N.; Dorofeev, G.A.; Karakishev, S.D.; Litvinov, V.S. Mössbauer spectroscopy study of iron-manganese austenite alloys. *Fiz. Met. Met.* **1973**, *36*, 666–668.
34. Renot, R.C.; Swartzendruber, L.J. Origin of Mössbauer linewidth in stainless steel. *AIP Conf. Proc.* **1973**, *10*, 1350–1353.
35. Srivastava, B.P.; Sarma, H.N.K.; Bhattacharya, D.L. Quadrupole splitting in deformed stainless steel. *Phys. Status Solidi A* **1972**, *10*, K117–K118. [CrossRef]
36. Dubiel, S.M.; Żukrowski, J. Distribution of Cr atoms in a strained and strain-relaxed Fe 89.15 Cr 10.75 alloy: A Mössbauer effect study. *Philos. Mag. Lett.* **2017**, *97*, 386–392. [CrossRef]
37. Mirkin, L.I. *X-ray Control of Engineering Materials*; Mashinostroenie: Moscow, Russia, 1979.
38. Dorofeev, G.A.; Elsukov, E.P. Thermodynamic Modeling of Mechanical Alloying in the Fe-Sn System. *Inorg. Mater.* **2000**, *36*, 1228–1234. [CrossRef]
39. Sagaradze, V.V.; Kataeva, N.V.; Zavalishin, V.A.; Kozlov, K.A.; Makarov, V.V.; Kuznetsov, A.R.; Rogozhkin, S.V.; Ustyugov, Y.M. Formation of low-temperature deformation-induced segregations of nickel in Fe–Ni-based austenitic alloys. *Philos. Mag.* **2020**, *100*, 1868–1879. [CrossRef]
40. Leslie, W.C.; Mieler, R.Z. The stabilization of austenite by closely spaced boundaries. *Trans. ASM* **1964**, *57*, 972–979.
41. Gruzin, P.L.; Rodionov, U.L.; Mikhailova, L.K.; Isphandiarov, G.G. Low-temperature isothermal alpha yields gamma transformations in ordered iron-nickel alloys. *Phys. Met. Metallograph.* **1977**, *44*, 178–179.
42. Hornbogen, E.; Meyer, W. Martensitische Umwadlung von mischkristallen mit kohärenten teilchen. *Acta Metall.* **1967**, *15*, 584–588. [CrossRef]
43. Shabashov, V.A.; Sagaradze, V.V.; Yurchikov, E.E.; Savel'eva, A.V. Mössbauer spectroscopy and electron microscopy studies of $\alpha\rightarrow\gamma$ transformation and stabilization of iron-nickel austenite. *Fiz. Met. Met.* **1977**, *44*, 1060–1070.
44. Shabashov, V.A.; Kozlov, K.A.; Zamatovskii, A.E.; Lyashkov, K.A.; Sagaradze, V.V.; Danilov, S.E. Short-Range Atomic Ordering Accelerated by Severe Plastic Deformation in FCC Invar Fe-Ni Alloys. *Phys. Met. Metallograph.* **2019**, *120*, 686–693. [CrossRef]
45. Kozlov, K.; Shabashov, V.A.; Zamatovskii, A.; Novikov, E.; Ustyugov, Y. Inversion of the Sign of the Short-Range Order as a Function of the Composition of Fe–Cr Alloys at Warm Severe Plastic Deformation and Electron Irradiation. *Metals* **2020**, *10*, 659. [CrossRef]
46. Shabashov, V.A.; Sagaradze, V.V.; Kozlov, K.A.; Ustyugov, Y.M. Atomic Order and Submicrostructure in Iron Alloys at Megaplastic Deformation. *Metals* **2018**, *8*, 995. [CrossRef]
47. Shabashov, V.A.; Kozlov, K.; Ustyugov, Y.; Zamatovskii, A.; Tolmachev, T.; Novikov, E. Mössbauer Analysis of Deformation-Induced Acceleration of Short-Range Concentration Separation in Fe-Cr Alloys—Effect of the Substitution Impurity: Sb and Au. *Metals* **2020**, *10*, 725. [CrossRef]
48. Goldschmidt, H.J. *Intersitial Alloys*; Butterworth-Heinemann: London, UK, 1967.

metals

MDPI

Article

Fe-X-B-Cu (X = Nb, NiZr) Alloys Produced by Mechanical Alloying: Influence of Milling Device

Albert Carrillo, Joan Saurina, Lluisa Escoda and Joan-Josep Suñol *

Department of Physics, University of Girona, C/Universitat de Girona 3, 17003 Girona, Spain; albert.carrillo.berlanga@gmail.com (A.C.); joan.saurina@udg.edu (J.S.); lluisa.escoda@udg.edu (L.E.)
* Correspondence: joanjosep.sunyol@udg.edu; Tel.: +34-872418757

Abstract: In this work, we analyze the influence of the milling device in the microstructural evolution of two Fe-X-B-Cu (X = Nb, NiZr) alloys produced by mechanical alloying (MA). The two milling devices are a planetary mil (P7) and a shaker mill (SPEX 8000). Microstructural analysis by X-ray diffraction detects the formation of a Fe rich solid solution. In the Fe-Nb-B-Cu alloy produced in the shaker mill also appears a Nb(B) minor phase, whereas in the Fe-NiZr-B-Cu alloy produced in the planetary mill, a minor disordered phase is formed. The comparative study regarding the energy transferred per unit of time in both devices determines that the shaker mill is more energetic. This fact explains that in the Fe-Nb-B-Cu alloy, Nb has not been introduced in the main Fe rich phase, whereas in the Fe-NiZr-B-Cu alloy milled in the shaker mill was formed the highly disordered phase. With regard to thermal analysis, the values of the apparent activation energies of the main crystallization process (above 200 kJ/mol) correspond to the crystalline growth of the nanocrystalline Fe rich phase.

Keywords: mechanical alloying; microstructure; Fe based alloys

Citation: Carrillo, A.; Saurina, J.; Escoda, L.; Suñol, J.-J. Fe-X-B-Cu (X = Nb, NiZr) Alloys Produced by Mechanical Alloying: Influence of Milling Device. *Metals* **2021**, *11*, 379. https://doi.org/10.3390/met11030379

Academic Editor: Tomasz Czujko

Received: 21 January 2021
Accepted: 21 February 2021
Published: 25 February 2021

1. Introduction

Nanocrystalline soft magnetic alloys have been obtained by rapid solidification techniques as melt-spinning, MS, (ribbon shape) or by mechanical alloying, MA (powder shape) [1,2]. Melt-spun ribbons are usually amorphous and further annealing is needed to develop nanocrystalline structure by primary crystallization [3]. Mechanical alloying favors the direct formation of powdered nanocrystalline alloys, that can be compacted/sintered to produce bulk pieces [4].

The nanocrystalline alloys can be classified based on the crystallographic structure of the nanocrystalline phase. The bcc Fe based alloys developed by Suzuki and coworkers are known as Nanoperm [5]. Nanoperm alloys, particularly nanocrystalline Fe-(Zr, Nb)-B-Cu alloys, have good magnetic properties: High effective permeability (~20,000) and low coercivity (<3 A/m) and core loss (<100 W/kg) at room temperature. There are soft magnetic offering reduced core losses over a wide range of applied frequencies. These alloys are an alternative for industrial (automotive, green energy) applications: High precision energy meters, absorber cores for suppression of motor bearing currents, high frequency current transformers, push-pull transformers for power transistors, and current sensors [3,4,6]. Due to magnetocaloric effect of the ferromagnetic to paramagnetic transition, these materials have also been of interest for magnetic refrigeration [7]. The composition is a key factor. It is known that an excessive content of B favors the formation of an amorphous phase, as it was found in $Fe_{75}Nb_{10}B_{15}$ [8], the increase of the coercivity and a decrease of the magnetization of saturation [9].

In the mechanical alloying process, the processing parameters influence the final microstructure and the functional properties. Milling parameters include the milling intensity, time and atmosphere, the ball-to-power weight ratio (BPR), the addition of a process control agent (PCA), and the filling factor of the vials. Some parameters, such as

PCA, are linked to grain refinement [10], but the majority has a high influence in the energy transferred to the powders [11].

In this work, two Nanoperm alloys ($Fe_{80}Nb_7B_{12}Cu_1$ and $Fe_{80}(NiZr)_7B_{12}Cu_1$) have been produced in two milling devices (shaker mill, planetary ball mill) to detect their influence on the nanocrystalline bcc Fe rich phase formation. In the SPEX 8000 shaker mill, the jar containing powders is agitated at a high frequency in a complex cycle that involves motion in three orthogonal directions (combining back-and-forth swings with short lateral movements, each end of the vial describing a figure like an eight). In the P7 planetary mill, two jars are arranged eccentrically on a platform (opposite direction of rotation between platform and jars). Furthermore, crystallographic microstructure (after 80 h MA) was discussed by energy transfer models.

2. Materials and Methods

Mechanical alloying has been performed in a Spex 800 shaker mill (Spex, Metuchen, NJ, USA) and in a Fritsch Pulverisette 7 planetary mill (Fritsch, Idar-Oberstein, Germany). The experiments were performed in jars sealed inside an Ar-filled glove box to prevent oxidation. The milling (10 g of powders) was performed in both devices with Cr-Ni steel jars (45 mL capacity) and balls (7 balls, 12 mm diameter) for 5, 20, and 80 h. The initial filling factor was close to one half. In SPEX, the information by the provider is that the number of cycles (rotational speed or frequency of rotation) is 875 rpm, whereas in the P7, the selected cycles of the platform, Ω, and jars, ω, are the same, 600 rpm. The R radius of the platform (measured from the geometric center of the jar position to the center of the platform) of the P7 is r_p = 7 cm and the internal radius of the jars r_j = 1.7 cm. The internal radius of the SPEX jar is 2.1 cm. The ball radius r_b is 0.6 cm in the Spex and 0.5 in the P7. The ball to powder weight ratio, BPR, used in both devices has been the same: 5.

The nominal composition of the two produced alloys were $Fe_{80}Nb_7B_{12}Cu_1$ and $Fe_{80}(NiZr)_7B_{12}Cu_1$ (at.%), labelled as A and B, respectively. Thus, depending on the mill device, the samples are labelled as: P7—A, SPEX—A, P7—B and SPEX—B. Elemental Fe, Nb, B, Cu powders, and NiZr compound (Sigma-Aldrich, Saint Louis, MO, USA) with high purity (<99.5%) and low particle size (Fe and Cu ~ 10 μm, B, B and Nb ~ 10 μm, NiZr < 100 μm) are mixed to obtain the desired composition. Likewise, NiZr compound was selected to avoid undesired oxidation of Zr.

The morphology, microstructure, and thermal behavior evolution as a function of milling time has been analyzed by scanning electron microscopy (SEM) coupled with compositional EDS spectroscopy microanalysis, X-Ray diffraction (XRD), and differential scanning calorimetry (DSC). SEM observations and EDS microanalysis were performed in a D500 device (Siemens/Bruker, Billerica, MA, USA); at least five micrographs/microanalyses for samples milled for 80 h. Thermal analysis was performed in a DSC822e calorimeter (Mettler-Toledo, Columbus, OH, USA); one experiment at 20 K/min for samples milled for 5 and 20 h, and four experiments for samples milled for 80 h. XRD experiments (before milling and after 5, 20, and 80 h of milling) were performed in a D8 Advance diffractometer (Siemens/Bruker, Billerica, MA, USA) using Cu-Kα radiation. XRD patterns analysis was performed by applying Rietveld refinement and the free software MAUD (Maud, Trento, Italy) [12].

3. Results and Discussion

3.1. Microstructure and Thermal Analysis

The XRD diffraction patterns of alloys A and B are given in Figures 1 and 2, respectively. The main component is elemental Fe in both alloys. Thus, as expected, the main phase (higher reflections intensity) in all the diffraction patterns is the bcc Fe phase. The reflection peaks of this phase are (110), (200), (211), and (220). The minor peaks before milling correspond to Cu, Nb, and NiZr compound. Initial elemental B was amorphous. These reflections disappear after milling. The minor elements were introduced in solid solution in the bcc Fe rich phase, or remains in the grain boundaries [8,9]. Regarding the shape of the peaks, its width has increased considerably with the milling time, phenomena associated to a decrease in the crystalline size. The parameter of the goodness of fit, GOF, obtained from Rietveld refinement is lower than 1.25 for all XRD patterns. Regarding the shape of the peaks, its width has increased considerably with the milling time that seems to indicate a decrease in the crystalline size.

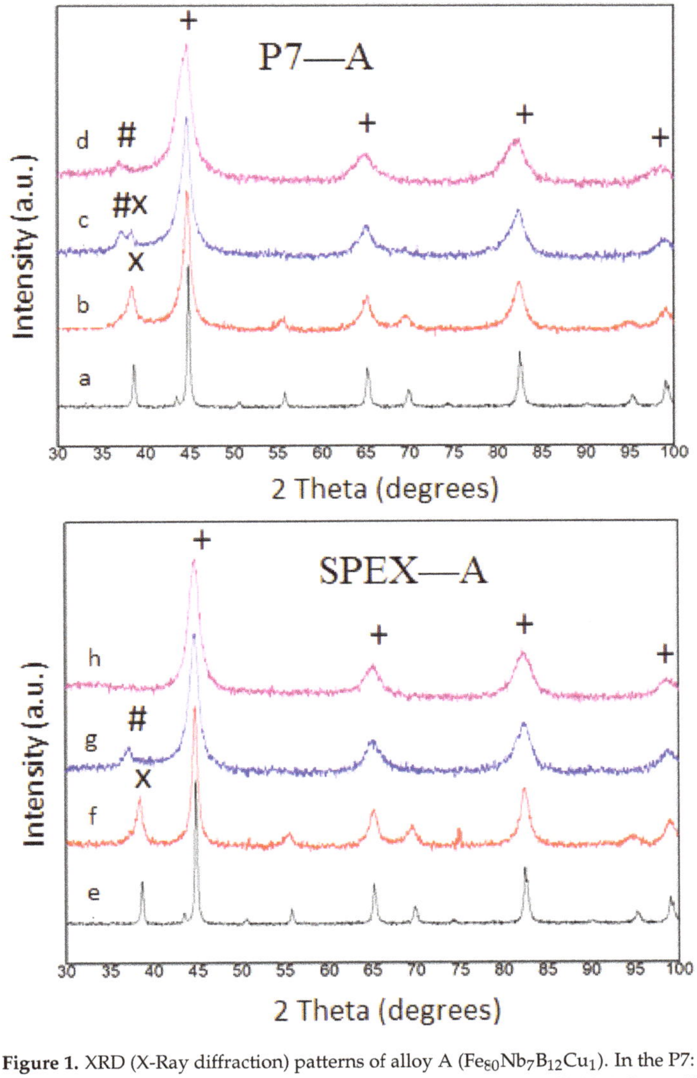

Figure 1. XRD (X-Ray diffraction) patterns of alloy A ($Fe_{80}Nb_7B_{12}Cu_1$). In the P7: (**a**) Before milling (0 h) and after milling ((**b**) 5 h, (**c**) 20 h, (**d**) 80 h); or in the Spex mill: (**e**) Before milling (0 h) and after milling ((**f**) 5 h, (**g**) 20 h, (**h**) 80 h). Bcc Fe rich solid solution (+), Nb phase (X), Nb(B) phase (#).

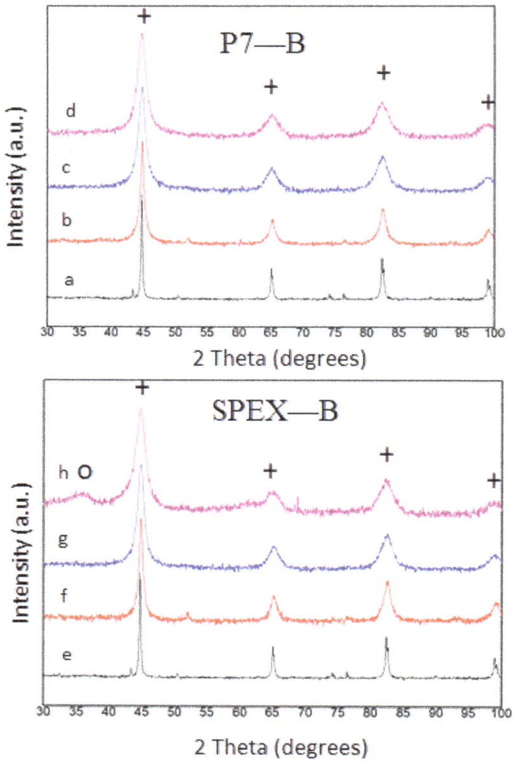

Figure 2. XRD diffraction patterns of alloy B ($Fe_{80}(NiZr)_7B_{12}Cu_1$). In the P7: (**a**) Before milling (0 h) and after milling ((**b**) 5 h, (**c**) 20 h, (**d**) 80 h); or in the Spex mill: (**e**) Before milling (0 h) and after milling ((**f**) 5 h, (**g**) 20 h, (**h**) 80 h). Bcc Fe rich solid solution (+), disordered phase (O).

If one focuses in the XRD patterns after 5 h of milling in alloy A (Figure 1a), the Nb peaks are still present. With the peak (110) of the Nb, it was verified that, at its lower angular reflection, we began to notice an asymmetry. The result of this asymmetry was clearer after 20 h of milling (in sample A), in which pattern appears a double peak in the place where before there was only one. Whereas in sample B, there only appears the new low angular phase. The proximity of the two peaks seems to indicate a mutual relationship. This new phase forcibly has a cell parameter slightly higher than that of the elemental Nb since it is shifted to lower angular positions. In short, asymmetry that began to be present at 5 h was an incipient peak of a new phase of Nb(B), which becomes more evident at 20 h. The creation of this phase is favored by the negative mixing enthalpy between Nb and B (−39 kJ/mol), indicating an exothermic process, which has also been detected in the scientific literature [13].

The formation of borides (undesired phases due to the associated loss of improved soft magnetic behavior) was not detected. After 80 h of milling (A alloy), in SPEX only the Fe rich bcc solid solution was detected, whereas a minor amount (2.1%) of the mechanically induced Nb(B) phase remains in alloy A milled in the P7 device.

The analysis of the XRD patterns of alloy B shows significative differences. The main phase is the bcc Fe rich phase and the reflections of minor phases disappears during milling. It should be remarked that the NiZr phase has not been detected after milling (probably the integration of Zr in the NiZr compound favors the integration of both elements in the main solid solution. From 80 h MA XRD diffraction patterns analysis, it was found that in one case (milling in P7) there is only the majority solid solution; while in the other case (milling in SPEX), a highly disordered phase appears (about 3.2%). This phase has been previously detected [10] in alloys with Zr and is probably due to an incipient amorphous Zr rich phase.

The main parameters derived from the Rietveld refinement (Maud software); the lattice parameter, a, the crystalline size, L, and the microstrain, ε, were given in Tables 1–3, respectively (as a function of the milling time). The values are calculated with the isotropic model. Thus, they are calculated from refinement weighted analysis of the reflections of the bcc Fe rich phase, and the error is associated to both the isotropic model and the mixed refinement of all parameters.

Table 1. Lattice parameter, a (Å), of the cubic bcc Fe main phase.

Sample/Milling Time	0 h	5 h	20 h	80 h
P7—A	2.8608 ± 0.0001	2.8681 ± 0.0002	2.8731 ± 0.0002	2.8832 ± 0.0003
SPEX—A	2.8608 ± 0.0001	2.8659 ± 0.0001	2.8706 ± 0.0002	2.8694 ± 0.0002
P7—B	2.8608 ± 0.0001	2.8664 ± 0.0001	2.8699 ± 0.0002	2.8705 ± 0.0002
SPEX—B	2.8608 ± 0.0001	2.8642 ± 0.0002	2.8672 ± 0.0002	2.8719 ± 0.0003

Table 2. Crystalline size, L (nm), of the cubic bcc Fe main phase.

Sample/Milling Time	0 h	5 h	20 h	80 h
P7—A	237 ± 8	19.9 ± 0.9	14 ± 1	9.9 0.4
SPEX—A	237 ± 8	36 ± 3	15 ± 1	15 ± 1
P7—B	237 ± 8	29.3 ± 0.9	15.7 ± 0.6	13.4 ± 0.5
SPEX—B	237 ± 8	16.5 ± 0.9	11.5 ± 0.7	9.5 ± 0.5

Table 3. Microstrain, ε (%), of the cubic bcc Fe main phase.

Sample/Milling Time	0 h	5 h	20 h	80 h
P7—A	0.028 ± 0.003	0.10 ± 0.02	0.41 ± 0.08	0.66 ± 0.05
SPEX—A	0.028 ± 0.003	0.22 ± 0.02	0.42 ± 0.03	0.61 ± 0.02
P7—B	0.028 ± 0.003	0.02 ± 0.01	0.47 ± 0.02	0.59 ± 0.06
SPEX—B	0.028 ± 0.003	0.19 ± 0.02	0.44 ± 0.04	0.40 ± 0.03

The milling process favors the increase of the lattice parameter. This effect is due to the introduction of mechanically induced crystallographic defects (vacancies, dislocations) and to the solid solution of minor elements in the bcc phase. The highest value, 2.8832 ± 0.0003 Å, corresponds to alloy P7—A milled for 80 h. The lower values are found in samples without an additional second minor phase. Probably, these low parameters are due to the no solid solution of Nb and Zr (a high amount of Nb and Zr atoms remains in the Nb(B) and disordered phases, respectively).

The milling process favors the decrease of the crystalline size and the increase of the microstrain. It should be remarked that low crystalline size is calculated in those alloys (P7—A and SPEX—B milled for 80 h) with a minor secondary phase. One explication is that the Nb(B) and the disordered Zr rich phases formation favors the intra-division of the highly deformed crystals. The increase on the microstrain is due to the mechanically induced crystallographic defects.

The density of dislocation (of the bcc phase) was calculated (see Table 4) from crystallographic parameters (from Tables 2 and 3) by applying the following equation [14]:

$$\rho = 2 \times \sqrt{3}\,(\varepsilon/Lb), \quad (1)$$

Table 4. Dislocations density, ρ (10^{15} m^{-2}), of the cubic bcc Fe main phase.

Sample/Milling Time	0 h	5 h	20 h	80 h
P7—A	0.0323 ± 0.003	1.10 ± 0.09	4.58 ± 0.08	9.72 ± 0.07
SPEX—A	0.0323 ± 0.003	1.12 ± 0.08	4.44 ± 0.03	6.22 ± 0.05
P7—B	0.0323 ± 0.003	0.07 ± 0.05	4.91 ± 0.12	7.32 ± 0.08
SPEX—B	0.0323 ± 0.003	1.10 ± 0.09	4.11 ± 0.09	7.08 ± 0.07

In this equation, b is the Burgers vector. For the bcc crystallographic phase, the direction of easy dislocations concentration is (1 1 1). In this case, the Burgers vector is:

$$b = a\ \sqrt{3}/2, \tag{2}$$

The general trend, as expected, it is the increase of the dislocations density as increasing the milling time. Higher values are calculated in those samples milled (80 h) in the planetary mill. The results are close to the expected limit of the dislocation density (plastic deformation) in bcc Fe (10^{16} m^{-2}) [15]. It should be remarked than XRD methods sometimes can provide higher values of the dislocations density than transmission electron microscopy (TEM) analysis [16]. TEM values are considered as more accurate due to direct local observation. Nevertheless, the milling process favors inhomogeneity in the crystallographic defects.

The morphology of the powders after 80 h of milling was checked with SEM micrographs (as shown in Figure 3). There are rounded particles with smooth shape.

Figure 3. SEM micrographs after 80 h milling of samples: (**a**) P7—A, (**b**) SPEX—A, (**c**) P7—B, and (**d**) SPEX—B.

In order to analyze the particle size and its distribution, the particle size of several micrographs of each sample have been measured. The results are shown in Figure 4. To facilitate the comparison, the results were fitted with a log-normal distribution with the mean, μ, and the shape parameter (square root of variance), σ, values are given in Table 5.

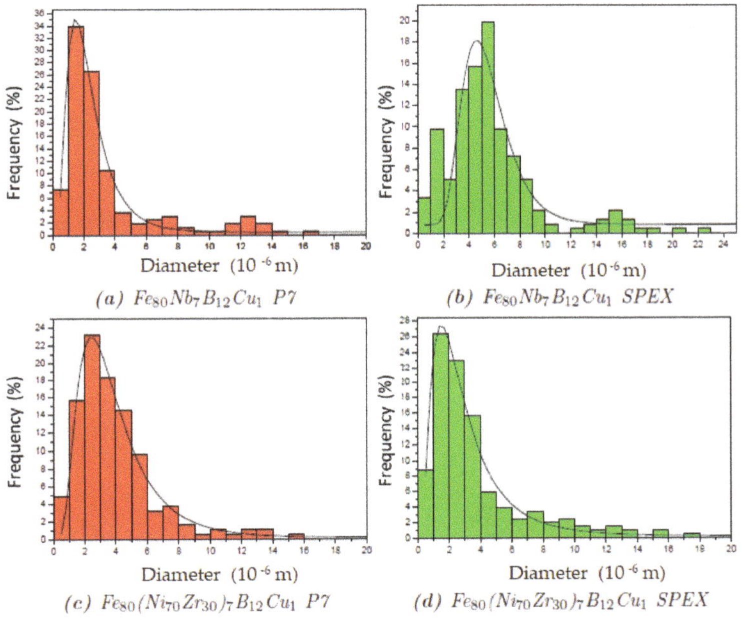

Figure 4. Particle size frequency distribution after 80 h milling of samples: (**a**) P7—A, (**b**) SPEX—A, (**c**) P7—B, and (**d**) SPEX—B.

Table 5. Parameters (mean, μ, and shape parameter, σ) of the log-normal distribution.

Sample	μ /10^{-6} m	σ /10^{-6} m
P7—A	2.4458	1.5367
SPEX—A	5.5279	1.9658
P7—B	3.9970	2.4863
SPEX—B	3.2053	2.5755

The higher mean value was found in sample SPEX—A, whereas the high shape parameter values are found in B samples. No correlation has been detected between these parameters (micrometric scale) and the XRD results (nanometric scale).

Likewise, compositional analysis was performed (80 h of milling) with EDS microanalysis. The measurements indicates that the contamination from the milling tools is lower than 1.5 ± 0.5 at.%; typical results coherent with those previously found [10,16]. Oxygen is also found, probably favored by the high surface area of the particles in contact with air after the milling process. Values are lower than 2.5 ± 0.5 at.%.

Additional thermal analysis has been performed from DSC scans. The analysis of the evolution of the thermal behavior has been carried out after 5, 20, and 80 h of milling at 10 K/min as the heating rate. The DSC scans (5 to 80 h of milling) of A and B samples are given in Figures 5 and 6, respectively. DSC after 80 h of milling were reported on ref. [17]. Here we discuss the process evolution and the kinetic analysis.

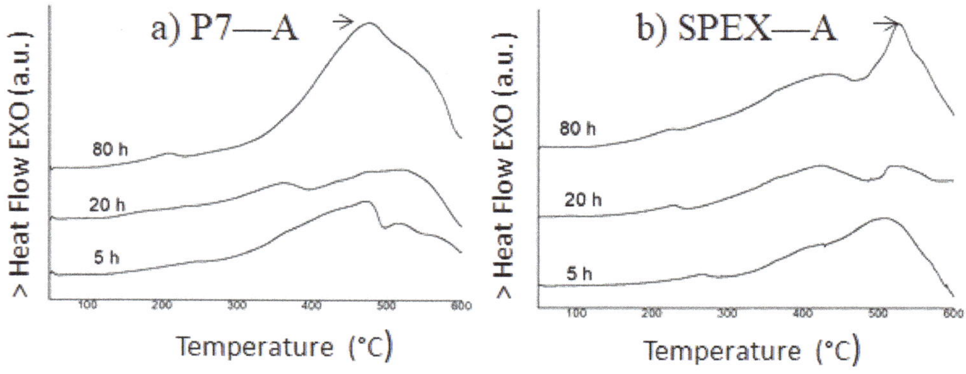

Figure 5. DSC (differential scanning calorimetry) scans of alloy A after milling (5, 20, 80 h) in the P7 (**a**) or in the Spex (**b**) mill.

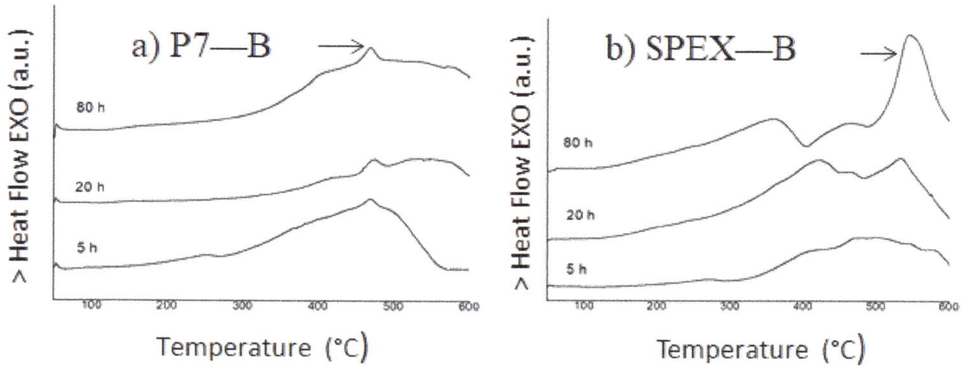

Figure 6. DSC scans of alloy B after milling (5, 20, 80 h) in the P7 (**a**) or in the Spex (**b**) mill.

In all scans, multiple exothermic processes are observed in wide intervals of temperature and with a certain degree of overlap. Exothermic processes at lowest temperature (below 400 °C) were associated with relaxation phenomena (caused by a decrease of the free volume after relaxation of the mechanically induced tensions at the micro and nanoscale). Furthermore, thermal treatment also favors the reduction of the crystallographic defects and local inhomogeneity by atomic diffusion. The exothermic processes at higher temperatures were associated with crystallization (nucleation and/or crystalline growth). This complex behavior is typical of nanocrystalline alloys produced by mechanical alloying [9,18]. In all DSC at 80 h, the main peak is remarked with an arrow. The peak temperature is higher ($T_p > 500$ °C) in samples milled with a shaker mill if compared with the temperature of the samples milled in the planetary mill ($T_p < 500$ °C). Probably, higher energetic milling favors higher internal local temperature and atomic diffusion [18].

On the other hand, the apparent activation of the crystallization processes (those remarked with an arrow) has also been calculated by applying Kissinger's linear method at heating rates of 10, 20, 30, and 40 K/min. Kissinger's method is based on the relationship between the peak temperature, T_p, and the scan rate, β. The linear fitting is shown in Figure 7, and the apparent activation energies are given in Table 6. The activation energy is determined from the slope of the linear fit, by applying the following relation (R is the universal gas constant):

$$E = -R/(1000\ T_p), \quad (3)$$

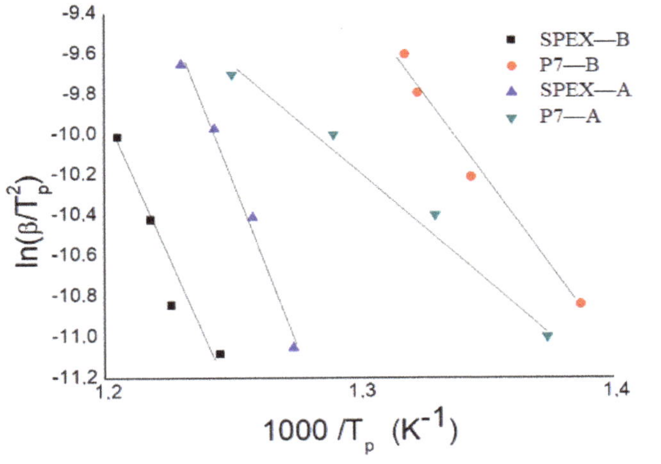

Figure 7. Kissinger linear fitting to calculate the activation energy of the crystallization process (experimental point for each sample are linked to peak temperatures of samples determined at 10, 20, 30, and 40 K/min from right to left respectively). Coefficients of correlation R-squared > 0.9.

Table 6. Apparent activation energy of the crystallization process.

Sample	E kJ mol^{-1}
P7—A	87 ± 11
SPEX—A	263 ± 21
P7—B	144 ± 13
SPEX—B	225 ± 44

The activation energies are higher in samples milled in the shaker mill, and a possible hypothesis is that it was favored by different milling energy transfer (higher energy can favor diffusion and the homogeneity of the alloys, stabilizing against crystalline growth). Local inhomogeneities can act as preferential sites for crystalline growth, reducing the activation energy. An alternative hypothesis is the content of oxygen, higher (~0.5–0.8 at.%) in powders milled in the shaker mill. More work will be provided to reach a definitive explanation. Likewise, the energy of activation values were characteristic for each type of reaction [19]. According to the bibliography, the activation energy for the crystal growth of Fe based nanocrystalline samples are around 140 kJ/mol [20] for pure iron. In samples containing additional elements, these energies can have higher/lower values. In Fe-Co-Nb-B alloys obtained by MA, energies between 238 and 261 kJ/mol have been calculated [21]. For amorphous samples in which a new phase must be nucleated, energies have values and are around 300 kJ/mol [22] or 365 kJ/mol [23], or even 425 and 550 kJ/mol in Fe-Nb-B alloys [24].

Thus, all of the analyzed crystallization process can be associated to the crystalline growth of the pre-existing Fe rich solid solution.

3.2. Mills Comparison

If we compare the phases detected by XRD after 80 h of milling, it is found that:

(a) In the composition with Nb milled in SPEX, only the Fe rich solid solution has been detected. Whereas, in the Nb alloy milled in P7, the major phase is the Fe rich bcc phase coexisting with a low amount of a Nb(B) phase.

(b) In the composition with Zr milled in SPEX, the Fe rich solid solution coexists with a minor disordered phase. Whereas, in the alloy milled in P7, this phase was not formed.

The results indicate that probably the milling is more energetic and efficient in SPEX. In one case, at 80 h in the P7, it was not possible to obtain a single phase (it would surely be necessary to increase the milling time to promote the Nb(B) phase disappearance and the integration of the Nb, via solid solution or in the grain boundaries). In the other, in the SPEX, the more energetic milling favors the formation of a disordered phase.

Likewise, from DSC scans we can state that the higher thermal stability (of the analyzed crystallization process) was found in the samples milled in the shaker mill. The criteria for a higher thermal stability of the nanocrystalline phase were high temperature of transformation and high activation energy.

A comparative study was performed to analyze the transfer of energy in both milling devices. This factor must influence the milling time needed to develop specific microstructural changes, such as nanocrystallinity. In the bibliography, there are multiple comparative studies between ball milling devices at the laboratory level. These are based on the comparison via experimental data and/or simulations. Overall, these studies usually conclude that the planetary mills were less energetic than the shaker mills as SPEX. Mathematic ab initio models have been applied to study MA process [25]. Moreover, experimental data to determine the speed of balls or rotation of milling vials [26]. These works are generally based on energy transferred during milling [27]. To make a comparison a useful parameter was the ratio between the power transmitted and the powder mass (P/m).

One of the problems associated with the energetic comparison of mills is the differences in the motion. Another question is that the factors involved are multiple and there are complex relationships between them: Rotation frequencies, number and characteristics of the balls, geometry vials, degree of container filling, powder mass, energy transfer, temperature, milling atmosphere, frequency of interactions, etc. Furthermore, real processes have a distribution of energies involved in collisions due to a distribution of ball velocities in the shocks [28]. Some of the studies carried out take into account the direct comparison of shaker Spex and planetary mills [19,29].

First, for the comparison between both devices, we apply the method described in ref. [29]. In order to evaluate the total energy transferred to the powder during milling, it is necessary to know the ball velocity. In the Spex device, the ball velocity given in ref. [29] is v = 2.5 m/s. To have an equivalent value in the P7, the following formula has been chosen to determine the speed, v, of ball impact [30].

$$v^2 = (r_p\,\Omega)^2 + (r_j - r_b)^2\,\omega^2\,(1 + (2\omega/\Omega)) \quad (4)$$

As introduced in the Section 2, in SPEX the cycles are 875 rpm, whereas in the P7 the selected number of cycles (jar, platform) is $\omega = \Omega = 600$ rpm. The P7 radius of the platform $r_p = 7$ cm, the ball radius r_b is 0.5 cm for P7; and the internal radius of the jar $r_j = 1.7$ cm. The calculated velocity is v = 0.79 m/s. Obviously, these values should be considered for devices comparison, and real energy transfers have a distribution of speeds. Other models provide different velocities, as an example, the expression of reference [31] provides a value of v = 2.04 m/s for P7.

Regarding the kinetic energy, E_c, transferred by a ball in the planetary mill, it can be calculated directly from the velocity. The calculated value is 0.002 J. The equivalent calculation in the shaker was calculated in ref. [29], and the kinetic energy is 0.026 J.

The normalized (by unit mass of powder) total energy (E/m) transferred at a given milling time, t, was calculated by taking into account the following expression [32]:

$$E/m = (nE_c t/m) \quad (5)$$

where n is the balls number. As shown in Table 7, the calculated values for 80 h of milling are $(E/m)_{SPEX} = 5193$ J/g and $(E/m)_{P7} = 3752$ J/g. Thus, milling is more energetic in the Spex device.

Table 7. Ball velocity, kinetic energy, and total energy of Spex and P7.

Mill	Ball Velocity m/s	Kinetic Energy J	(E/m) J/g
P7	0.79	0.002	3752
SPEX	2.5	0.026	5193

The applied model considers that the frequency of impacts (events frequency) is equivalent in both devices. This factor is inversely proportional to the time between shock events on milling. The sliding factor influences the time it takes for the balls to separate from the container wall, the takeoff time, t_t. The increase in this time reduces the frequency of collisions. It is also necessary to determine a flight time of the ball before impact. Obviously, there is a distribution of speeds as well as energy and time. However, it should be taken into account that we need to stablish an easy way to compare both devices with characteristic mean parameters.

The ball detaches from the container wall when the value of the contact force is zero. The takeoff time can be estimated by applying the following expression [33]:

$$t_{t\,=}\ (1/(\Omega + \omega))\ \arccos\left(-(\omega^2\ r/\Omega^2\ R)\right), \tag{6}$$

In the P7 device this time is $t_t = 0.076$ s. Likewise, the flight time t_f (between detachment and collision) is also linked to the period (inverse of the frequency) in planetary mills. Thus, the period is higher in P7. The P7 flight time is determined by using the relationship given in Figure 8 [19]. The calculated time is $t_f = 0.013$ s.

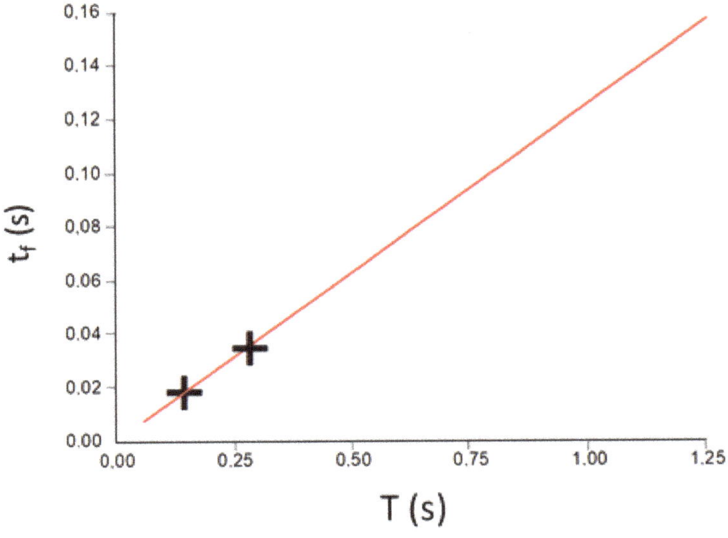

Figure 8. Flight time, t_f, as a function of the period T, (+, this work; red line, [19]).

Consequently, the total time [19], its sum, is $t_t + t_f$ (P7) = 0.089 s. In the SPEX device, the bibliography data were used: $t_t = 0.0073$ s and $t_f = 0.0269$ s that drive to $t_t + t_f$ (SPEX) = 0.0342 s. As the total time is higher in the planetary mill, the events frequency (for unit time) is higher in the shaker mill.

Thus, from the time between collisions, the shaker mill is also more effective. Likewise, the results obtained in this study cannot be directly based in this total time because the final microstructure varies (sometimes appears a second minor phase).

Therefore, SPEX mill is more energetically efficient than the P7 planetary mill (in the processing conditions of our study).

On the other hand, it is possible to calculate an equivalent milling time between different mills [34,35]. The calculation of this equivalent time is based on the assumption that the power transferred per unit of mass (P/m) is directly proportional to the time so that in two different devices the powder achieves the same final microstructure. It can be determined by applying the following equation:

$$(P/m)_{SPEX} = \varphi\ (P(\Omega_o)/m)_{P7}\ (\Omega/\Omega_o)^3, \tag{7}$$

This approach has been previously applied in FeNb and FeZr alloys [35]. In our study, the cycles ratio in the P7 is 1 ($\Omega = \Omega_o$= 600 rpm). From the previous calculations, we can calculate φ that is a proportionality factor, φ = 1.384. Thus, equation 6 is reduced to:

$$(P/m)_{SPEX} = \varphi\ (P(\Omega_o)/m)_{P7}, \tag{8}$$

The equivalent milling time, t_{eq}, can be calculated by applying [19]:

$$t_{eq}= ((P/m)_{SPEX}/(P(\Omega_o)/m)_{P7})\ t, \tag{9}$$

It has been estimated (from Table 7 data) that the time required in the P7 for the same energy transfer is 1.384 higher than for SPEX. Obviously, there are many factors modifying the consequences of this equivalent time. Sometimes the MA process is not energetic enough to obtain a concrete microstructure. For example, by modifying the milling conditions, the achieved microstructure can be nanocrystalline or amorphous [36]. Other possibilities are that the energy transfer facilitates the extension of the solubility limits or the formation of intermetallic compounds [24,37]. In alloy A, the energy/power transfer is not enough to obtain only the bcc Fe rich solid solution in the planetary mill (Nb(B) phase remains after 80 h of milling), and in alloy B the high energy/power transfer in shaker mill favors the formation of an incipient amorphous phase. Additional experiments, with alloys of similar composition and/or by repeating production process will confirm the general trend found in this work.

4. Conclusions

Two alloys, $Fe_{80}Nb_7B_{12}Cu_1$ (A) and $Fe_{80}(NiZr)_7B_{12}Cu_1$ (B), have been produced in two milling devices, a shaker mill (SPEX) and a planetary mill (P7). In the sample A milled in the shaker device and in sample B milled in the planetary mill only an Fe rich solid solution was detected by XRD after 80 h of MA. Instead, in the sample A milled in the planetary mill a low percentage of a Nb(B) phase was also detected; whereas in the sample B shaker milled the Fe rich solid solution coexist with a minor disordered phase. Additional thermal analysis permits us to establish that the analyzed crystallization process is due to crystalline growth, and the thermal stability is higher in samples milled in the shaker mill (higher temperature of crystallization and activation energy).

The results indicate that probably the milling is more energetic and efficient in the shaker mill. The equivalent time required in the planetary mill for the same energy transfer is 1.384 higher. In alloy A, the energy/power transfer is not enough to obtain only the bcc Fe rich solid solution in the planetary mill (a minor Nb(B) phase remains), and in sample B the high energy/power transfer in shaker mill favors the formation of a disordered Zr rich phase.

Author Contributions: Methodology, L.E., J.S.; formal analysis, A.C.; investigation, A.C.; writing—original draft preparation, J.-J.S. All authors have read and agreed to the published version of the manuscript.

Funding: This research was funded by Spanish MINECO project MAT2016-75967-P and UdG project PONTOS-2020/01.

Institutional Review Board Statement: Not applicable.

Informed Consent Statement: Not applicable.

Data Availability Statement: The data presented in this study are available on request from the corresponding author.

Acknowledgments: The authors agree the Technical Research Facilities of the University of Girona.

Conflicts of Interest: The authors declare no conflict of interest.

References

1. Kwon, S.; Kim, S.; Yim, H.; Kang, K.H.; Yoon, C.S. High saturation magnetic flux density of novel nanocrystalline core annealed under magnetic field. *J. Alloys Comp.* **2020**, *826*, 154136. [CrossRef]
2. Alleg, S.; Ibrir, M.; Fenineche, N.E.; Azzaza, S.; Bensalem, R.; Suñol, J.J. Magnetic and structural characterization of the mechanically alloyed $Fe_{75}Si_{15}B_{10}$ powders. *J. Alloys Comp.* **2010**, *494*, 109–115. [CrossRef]
3. Parsons, R.; Zang, B.; Onodera, K.; Kishimoto, H.; Kato, A.; Suzuki, K. Soft magnetic properties of rapidly-annealed nanocrystalline Fe-Nb-B-(Cu) alloys. *J. Alloys Comp.* **2017**, *723*, 408–417. [CrossRef]
4. Suryanarayana, C. Mechanical alloying and milling. *Prog. Mater. Sci.* **2001**, *46*, 1–184. [CrossRef]
5. Suzuki, K.; Kataoka, N.; Inoue, A.; Makino, A.; Masumoto, T. High saturation magnetization and soft magnetic properties of bcc Fe-Zr-B alloys with ultrafine grain structures. *Mater. Trans.* **1990**, *31*, 743–746. [CrossRef]
6. Fan, X.; Tang, Y.; Shi, Z.; Jiang, M.; Shen, B. The effect of Ni addition on microstructure and soft magnetic properties of FeCoZrBCu nanocrystalline alloys. *AIP Adv.* **2017**, *7*, 056107. [CrossRef]
7. Alleg, S.; Chabi, T.; Bensebaa, N.; Saurina, J.; Escoda, L.; Hlil, E.K.; Suñol, J.J. Investigation on the critical behavior, magnetocaloric effect and hyperfine structure in the $Fe_{72}Nb_8B_{20}$ powders. *Materials* **2020**, *13*, 4476. [CrossRef] [PubMed]
8. Ipus, J.J.; Blázquez, J.S.; Franco, V.; Lozano-Pérez, S.; Conde, A. Role of the starting phase of boron on the mechanical alloying of FeNbB composition. *J. Alloys Comp.* **2013**, *553*, 119–124. [CrossRef]
9. Chabi, T.; Bensebaa, N.; Alleg, S.; Azzaza, S.; Suñol, J.J.; Hlil, E.K. Effect of the boron content on the amorphization process and magnetic properties of the mechanically alloyed $Fe_{92-x}Nb_8B_x$ powders. *J. Supercond. Novel Magn.* **2019**, *32*, 893–901. [CrossRef]
10. Pilar, M.; Suñol, J.J.; Bonastre, J.; Escoda, L. Influence of process control agents in the development of a metastable Fe-Zr based alloy. *J. Non-Cryst. Solids* **2007**, *353*, 848–850. [CrossRef]
11. Djilali, Z.; Said, B. Study of the ball milling device for synthesizing nanocrystalline powder. *J. Nano Res.* **2017**, *47*, 60–70. [CrossRef]
12. Lutterotti, L.; Matthies, S.; Wenk, H.R. Quantitative phase analysis. *Newsl. CPD* **1999**, *21*, 14–15. [CrossRef]
13. Alleg, S.; Kartout, S.; Ibrir, M.; Azzaza, S.; Fenineche, N.E.; Suñol, J.J. Magnetic, structural and therml properties of the Finemet-type powders prepared by mechanical alloying. *J. Phys. Chem. Solids* **2013**, *74*, 550–557. [CrossRef]
14. Eckert, J.; Holzer, J.C.; Krill, C.E.; Johnson, W.L. Structural and thermodynamic properties of nanocrystalline fcc metals prepared by mechanical attrition. *J. Mater. Res.* **1992**, *7*, 1751–1761. [CrossRef]
15. Ibn Gharsallah, H.; Azabou, M.; Escoda, L.; Suñol, J.J.; López, I.; Llorca-Isern, N. The magnetic and structural properties of $(Fe_{75}Al_{25})_{100-x}B_x$alloys prepared by mechanical alloying. *J. Alloys Comp.* **2017**, *729*, 776–786. [CrossRef]
16. Takebayashi, T.; Kunieda, T.; Yoshinaga, N.; Ushioda, K.; Ogata, S. Comparison of the dislocation density in martensitic steels evaluated by some X-ray diffraction methods. *ISIJ Inter.* **2010**, *50*, 875–882. [CrossRef]
17. Carrillo, A.; Escoda, L.; Saurina, J.; Suñol, J.J. Structural and magnetic behavior of Fe(Nb,Zr) rich alloys produced by mechanical alloying. *AIP Adv.* **2018**, *8*, 047704. [CrossRef]
18. Abdoli, H.; Ghanbari, H.; Baghshahi, S. Thermal stability of nanostructured aluminium powder synthesized by high energy milling. *Mater. Sci. Eng. A* **2011**, *528*, 6702–6707. [CrossRef]
19. Ipus, J.J. Microstructure and Thermomagnetic Behavior of (FeCo)(NbZr)(GeB) Alloys Prepared by Milling. Ph.D. Thesis, Universidad de Sevilla, Sevilla, Spain, 2009.
20. Ye, F.; Lu, K. Crystallization kinetics of Al-La-Ni amorphous alloy. *J. Non-Cryst. Solids* **2000**, *262*, 228–235. [CrossRef]
21. Liu, Y.; Chang, I.; Lees, M. Thermodynamic and magnetic properties of multicomponent $(Fe,Ni)_{70}Zr_{10}B_{20}$ amorphous alloy powders made by mechanical alloying. *J. Mater. Sci. Eng.* **2001**, *304–306*, 992–996. [CrossRef]
22. Suñol, J.J.; Güell, J.M.; Bonastre, J.; Alleg, S. Structural study of nanocrystalline Fe-Co-Nb-B alloys prepared by mechanical alloying. *J. Alloys Comp.* **2009**, *483*, 604–607. [CrossRef]
23. Krakhmalev, P.V.; Yi, D.; Nyborg, L.; Yao, Y. Isothermal grain growth in mechanically alloyed nanostructured $Fe_{80}Ti_8B_{12}$alloy. *Mat. Lett.* **2003**, *57*, 3671–3675. [CrossRef]
24. Johnson, F.; Hughes, P.; Gallagher, R.; Harris, V. structure and magnetic properties of new FeCo-based nanocrystalline ferromagnets. *IEEE Trans. Magn.* **2001**, *37*, 2261–2263. [CrossRef]
25. Ipus, J.J.; Blázquez, J.S.; Conde, C.F.; Borrego, J.M.; Franco, V.; Lozano-Pérez, S.; Conde, A. Relationship between mechanical amorphization and boron integration during processing of FeNbB alloys. *Intermetallics* **2014**, *49*, 98–105. [CrossRef]
26. Chattopadhyay, P.; Manna, I.; Talapatra, S.; Pabi, S. A mathematical analysis of milling mechanics in a planetary ball mill. *Mater. Chem. Phys.* **2001**, *68*, 85–94. [CrossRef]

27. Basset, D.; Matteazi, P.; Miani, F. Measuring the impact velocities of balls in high energy mills. *Mater. Sci. Eng. A* **1994**, *174*, 71–74. [CrossRef]
28. Concas, A.; Pisu, M.; Lai, N.; Cao, G. Modelling of comminution in processes in Spex Mixer/Mill. *Chem. Eng. Sci.* **2006**, *61*, 3746–3760. [CrossRef]
29. Dastanpoor, E.; Enayati, M.H. Effect of milling intensity on mechanical alloying of Cu-Zr-Al system. *Ind. J. Eng. Mater. Sci.* **2015**, *22*, 521–526.
30. Gheisari, K.; Oh, J.T.; Javadpour, S.; Ghaffari, M.J. The effect of milling speed on the structural properties of mechanically alloyed Fe-45% Ni powders. *J. Alloys Comp.* **2009**, *472*, 416–420. [CrossRef]
31. Magini, M. The role of energy transfer in mechanical alloying powder processing. *Mater. Sci. Forum* **2009**, *88*, 121–128. [CrossRef]
32. Singh, S.; Godkhindi, M.M.; Krishmarao, R.V.; Murty, B.S. Effect of milling energy on mechanical activation of (Mo+Si_3N_4) powders during the synthesis of Si_3N_4-$MoSi_2$in situ composites. *J. Euro. Ceram. Soc.* **2009**, *29*, 2069–2077. [CrossRef]
33. Karthik, B.; Sai Gautam, G.; Karthikeyan, N.R.; Murty, B.S. Analysis of mechanical milling in Sismoloyer: An energy modeling approach. *Metall. Mater. Trans. A Phys. Metall. Mater. Sci.* **2012**, *43*, 1323–1327. [CrossRef]
34. Ipus, J.J.; Blázquez, J.S.; Franco, V.; Millán, M.; Conde, A.; Oleszak, D.; Kulik, T. An equivalent time approach for scaling the mechanical alloying process. *Intermetallics* **2008**, *16*, 470–478. [CrossRef]
35. Blázquez, J.S.; Ipus, J.J.; Conde, C.F.; Conde, A. Comparison of equivalent ball milling process on $Fe_{70}Zr_{30}$ and $Fe_{70}Nb_{30}$. *J. Alloys Comp.* **2012**, *536*, 9–12. [CrossRef]
36. Martin, G.; Gaffet, E. Mechanical alloying: Far from equilibrium phase transitions. *Colloq. Phys.* **1990**, *51*, 71–77. [CrossRef]
37. Nova, K.; Novak, P.; Prusa, F.; Kopecek, J.; Cech, J. Synthesis of intermetallics in Fe-Al-Si system by mechanical alloying. *Metals* **2019**, *9*, 20. [CrossRef]

metals MDPI

Review

Mechanical Alloying of Elemental Powders into Nanocrystalline (NC) Fe-Cr Alloys: Remarkable Oxidation Resistance of NC Alloys

R. K. Singh Raman [1,2]

1 Department of Mechanical and Aerospace Engineering, Monash University, Clayton, VIC 3800, Australia; raman.singh@monash.edu
2 Department of Chemical Engineering, Monash University, Clayton, VIC 3800, Australia

Abstract: Mechanical alloying is among the few cost effective techniques for synthesizing nanocrystalline alloy powders. This article reviews mechanical alloying or ball-milling of (NC) powders of Fe-Cr alloys of different compositions, and the remarkable oxidation resistance of the NC alloy. The article also reviews challenges in thermal processing of the mechanically alloyed powders (such as compaction into monolithic mass) and means to overcome the challenges.

Keywords: nanocrystalline (NC) alloy; microcrystalline (MC) alloy; mechanical alloying; ball-milling; oxidation resistance

Citation: Singh Raman, R.K. Mechanical Alloying of Elemental Powders into Nanocrystalline (NC) Fe-Cr Alloys: Remarkable Oxidation Resistance of NC Alloys. *Metals* **2021**, *11*, 695. https://doi.org/10.3390/met11050695

Academic Editors: Tadeusz Kulik and Javier S. Blázquez Gámez

Received: 12 March 2021
Accepted: 22 April 2021
Published: 23 April 2021

1. Introduction

Nanocrystalline (NC) structure of alloys (instead of their traditional microcrystalline (MC) counterparts) is reported to remarkably improve some of the properties, such as strength and corrosion resistance [1–4]. Mechanical alloying of powders of the constituent elements is among the most commonly employed approaches for producing NC alloys [2,4–6]. This article reviews challenges in consolidation of mechanically alloyed Fe-Cr powders into monolithic solids and their sintering for achieving close-to-theoretical density, and circumvention of the challenges. The article also presents examples of improvements in oxidation/corrosion resistance properties of a few alloy systems due to the NC structure, and their mechanistic explanations.

2. Mechanical Alloying for Nanocrystallinity

A few processing techniques have been employed for synthesis of nanocrystalline (NC) alloys, such as electrodeposition [7], mechanical alloying [2,4–6], severe plastic deformation [8], sputtering, electron beam evaporation, pulse laser ablation [9], gas condensation [9,10], and sol-gel [11]. Some of these techniques (viz., sputtering, electron beam evaporation, pulse laser ablation, gas condensation and sol-gel) have limitations of producing only thin films or tiny samples of NC materials. Several critical applications, such as mechanical or corrosion testing, require bulk monolithic samples. For synthesis of bulk NC materials, mechanical alloying [2,4–6] and electrodeposition [7] have been widely employed.

A few metals [7], and Ni-Fe and Ni-Co alloys [5,6,12,13] were among the early NC metallic materials synthesized by pulsed electrodeposition. However, processing of NC metallic materials by pulsed electrodeposition requires promotion of nucleation and suppression of growth of the depositing grains, and, in order to bias nucleation, organic additives are added to the electrolyte used for electrodeposition [7]. A post-synthesis thermal decomposition is employed for removal of these additives. However, often, the organic additives still remain in the electrodeposited NC material and deteriorate its properties (such as causing embrittlement). In contrast, NC metal and alloy powders synthesized by mechanical alloying technique are generally free from such artefacts [2,4–6]. For critical applications such as mechanical or corrosion testing, the mechanically alloyed powders need to be compacted into monolithic mass. Depending on the thermal and mechanical characteristics of NC materials, various techniques have been employed for their compaction [14], such as hot compaction [15], high pressure/lower temperature compaction, in-situ consolidation [16], hot isostatic pressing [17], and explosive compaction [18]. Often, the consolidated material is subjected to sintering in order to achieve the theoretical density. The thermal stability of NC material governs the choice of sintering temperature and time. Examples of mechanical alloying of elemental powders into NC alloy powders of a few systems, their consolidation and sintering, and the associated challenges and circumvention are discussed in the succeeding sections. The role of the NC structure in remarkably influencing oxidation resistance is also discussed.

3. Mechanical Alloying of Fe-Cr Alloy Powders and Their Consolidation

Powders of Fe and Cr (purity: 99.9%, particle size: <10 μm) were mechanical alloyed into Fe-Cr alloy powders with different Cr contents (2, 5, 10, and 20 wt%) for the purpose of investigation of the role of nanocrystalline (NC) structure of Fe-Cr alloys in their oxidation/corrosion resistance [19]. NC Fe-Cr alloy powders were synthesized under an inert environment of toluene, using an air-cooled planetary ball-mill (300 rpm), tungsten carbide balls, and a powder-to-ball weight ratio of 1:10. The ball milling was continued for 20 h, and the crystallite grain size was determined intermittently, from the broadening of X-ray diffraction (XRD) peaks and the Scherrer equation [20]. Figure 1 shows the decrease in the grain size with increasing milling duration and confirms the grain size to have stabilized well and truly in the nanocrystalline regime (~20 nm) for all the powders, after 15–20 h of mechanical alloying [21]. It should be noted that the indirect determination of nanocrystalline grain size by XRD has been confirmed by direct measurements by transmission electron microscopy [21].

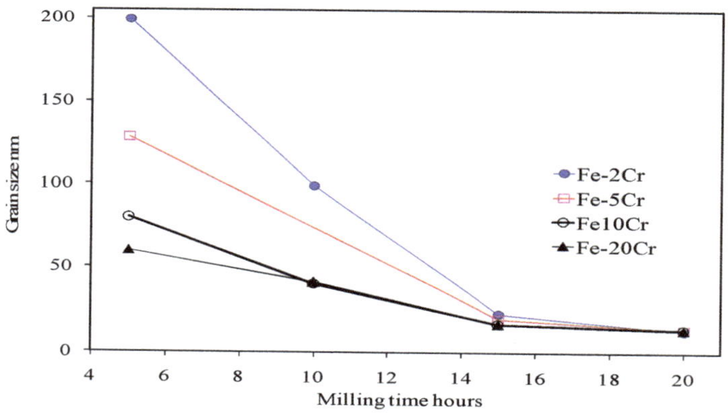

Figure 1. Decrease in the crystallite grain size with increasing duration of milling of elemental powders into Fe-Cr alloy powders with different Cr contents [21].

As mentioned earlier, for critical applications such as generating specimens for mechanical or corrosion testing, the mechanically alloyed powders need to be compacted into monolithic mass. However, it may be a non-trivial task to accomplish compaction of powders of NC Fe-Cr alloys, because the crystal structure of Fe and Fe-Cr alloys is body centred cubic (BCC), which resist plastic deformation and have high hardness. Therefore, the plastic deformation and material flow that are necessary for an effective compaction required for consolidation of Fe and Fe-Cr alloys necessitate application of high pressures and high temperatures. Seigel [22] reported that for plastic deformation of NC (10 nm) Fe (that has a hardness of 10 GPa), the pressure needs to be greater than the yield strength, i.e., 3.5 GPa (~1/3rd of hardness), that can be prohibitively high. Therefore, to accomplish compaction at moderate loads, it becomes necessary to elevate the compaction temperature in order to enhance plastic flow. However, the compaction temperature cannot be too high because that will cause excessive grain growth and loss of nanocrystallinity. Therefore, it becomes essential to identify the temperature-time window for accomplishing compaction without allowing excessive grain growth and loss of NC structure.

Malow et al. [23] successfully compacted NC Fe to ~100% of the theoretical density, by employing a pressure of 2.7 GPa at 475 °C. However, when the NC Fe-10Cr alloy powder that was prepared as described earlier was compacted at 2 GPa and 515 °C, only 92% of the theoretical density could be achieved, indicating the need for increasing the compaction temperature. To identify an appropriate temperature-time window for softening of hardened NC Fe-Cr alloy powders but avoiding the excessive grain growth, the mechanically alloyed powders produced as shown in Figure 1 [21] were subjected to annealing at 500, 600, and 700 °C in a reducing environment for different durations (3–120 min). The average grain size was determined after intermittent durations of annealing (again, from the broadening of XRD peaks and the Scherrer equation [20]). Figure 2 shows a typical set of XRD spectra for Fe-10Cr alloy mechanically alloyed for 3–120 min at 500 °C. The peaks are seen to broaden with increasing time of annealing because of the increasing grain growth.

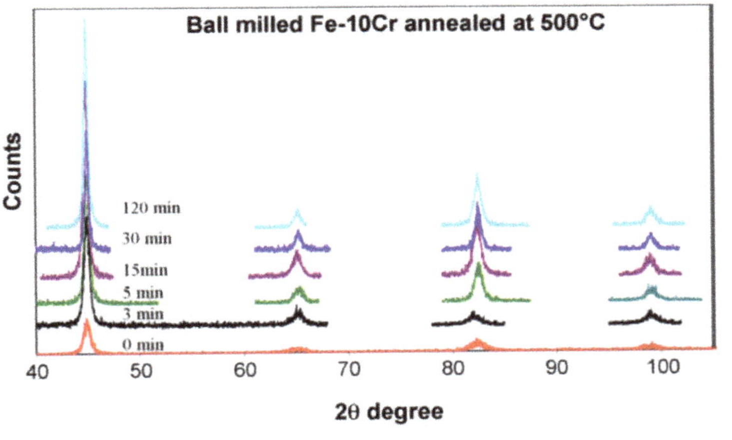

Figure 2. XRD spectra for mechanically alloyed Fe-10Cr alloy annealed for different durations (0–120 min) at 500 °C ([24], reproduced with copyright clearance).

Figure 3 shows the grain growth of Fe-10Cr alloy powder during annealing at 500, 600, and 700 °C for durations up to 120 min. Though the grains grow rapidly during the initial stage of annealing at each temperature, the intensity of growth is faster at higher temperatures. Grain growth plateaued after 15 min at 500 °C, whereas grains continued to grow rapidly at 600 and 700 °C. In fact, at 700 °C, the grain growth was so rapid that after 60 min the grains tended to grow beyond the nanocrystalline regime. The grain growth took off after 90 min also at 600 °C. Figure 4 compares the grain growth data for the mechanically alloyed Fe-10Cr NC alloy powder extracted from Figure 3, which are compared with those reported in the literature for ball-milled powders of NC Fe and an Fe-Al alloy [9,23–28]. There are considerable variabilities, particularly at temperatures above 400 °C. Such variabilities are attributed largely to the impurities in the powders. Some impurities such as oxide particles pin the grain boundaries and impede grain growth, and therefore powders with such impurities resist grain growth over higher temperature regimes. So, the purer the NC powder, the greater is its susceptibility to grain growth, and the lower is the temperature at which the rapid grain growth starts. The grain growth impeding due to impurities is more readily offered by alloying elements with greater thermodynamic susceptibility to form oxides, for example the Fe-Al alloy in Figure 4 shows resistance to grain growth over a higher temperature regime [28]. In fact, this effect has been successfully exploited for increasing the temperature regime for stability of NC alloys, as will be discussed subsequently.

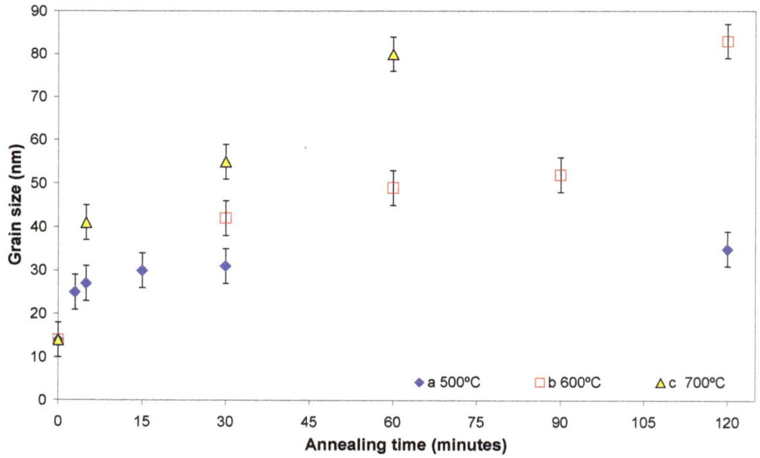

Figure 3. Grain growth of mechanically alloyed Fe-10Cr nanocrystalline alloy powder during annealing at 500 °C (**a**), 600 °C (**b**) and 700 °C (**c**) for different durations up to 120 min ([24], reproduced with copyright clearance].

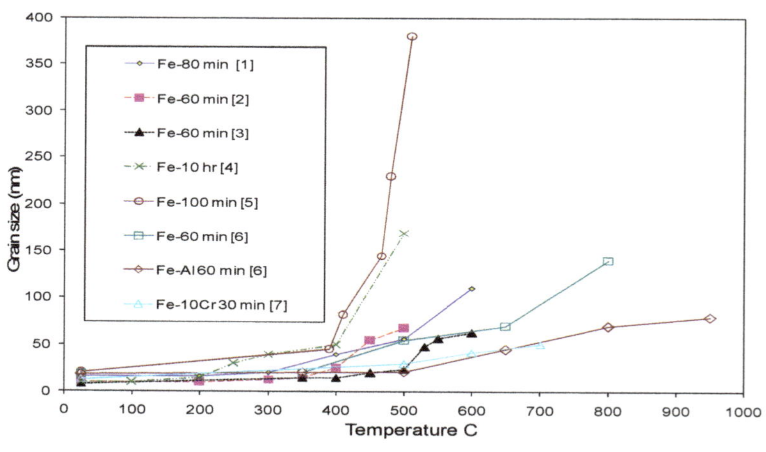

Figure 4. Grain growth data in literature for ball-milled nanocrystalline Fe and two Fe-alloy powders during annealing at different temperatures for different durations: [1] *NanoStructured Materials, 6 (1995) 421 (Ref* [25]), [2] *NanoStructured Materials, 12 (1999) 685 (Ref* [26]), [3] *Acta Mater, 45 (1997) 2177 (Ref* [23]), [4] *Progress in Materials Science, 33 (1989) 223 (Ref* [9]), [5] *Mater. Sci. Forum, 343 (2000) 683 (Ref* [27]), [6] *NanoStructured Materials 9 (1997) 71 (Ref* [28]), and [7] *Mater Sci Engineering-A, 494 (2008) 253 (Ref* [24]).

On the basis of the grain growth data in Figure 3, a time-temperature window of 600 °C and 30 min was selected for improving ductility and flow of mechanically alloyed NC powder of Fe-10Cr alloy for successful compaction of the powder into monolithic discs (under a uniaxial pressure of 2.7 GPa). The consolidated discs are seen in Figure 5. The density of the discs was determined to be 98% of the theoretical density. The compacted discs were sintered at 600 °C for 1 h to improve the density to ~99%. Again, grain growth data in Figure 3 were the basis for the choice of sintering time and temperature. Grain size of the sintered discs (52 nm) was very much in the nanocrystalline regime.

4. Oxidation Resistance of Mechanically Alloyed NC Fe-Cr Alloys

Fe-alloys with sufficient contents of specific alloying elements (namely Cr, Al and Si) develop a continuous surface layer of oxides of such alloying elements (i.e., Cr_2O_3, Al_2O_3 or SiO_2) that remarkably retards oxidation rate of such alloys. Defect contents of these oxides are remarkably low [29]. As a result, the counter diffusivity of elements involved in oxidation is insignificantly low through the surface layer of such oxides, hence the remarkable oxidation resistance of such alloys. Enabling the development of a continuous surface layer of these protective oxides is the basis for corrosion resistance of alloys. Most common examples are stainless steels (i.e., Fe-Cr alloys) and Al-alloys that readily develop robust surface barrier layers of Cr_2O_3 and Al_2O_3 respectively, and hence, such alloys possess excellent oxidation/corrosion resistance [29].

The development of a full-fledged continuous external layer of chromia, alumina or silica requires availability of a sufficient amount of the specific alloying elements (viz., Cr, Al or Si) at the surface [29]. The critical concentration of such alloying content for a given alloy can be determined by Wagner's calculations [30] (e.g., stainless steels possess 18–20 wt% Cr [3]). The critical concentration is governed directly by the diffusivity of the element (and the factors that influence diffusivity, e.g., temperature). It is well-known that at moderately elevated temperatures, grain boundary diffusivity far exceeds the lattice/bulk diffusivity [31]. The temperature range for the Fe-Cr system is 300–400 °C (as shown in Table 1). Therefore, it is obvious that diffusivity will be higher in the case of a nanocrystalline (NC) alloy (that possess remarkably high grain boundary volume fraction [9]) than its microcrystalline counterpart. Indeed, Wang et al. [31] reported the Cr diffusion in a NC iron to be four orders of magnitude greater than that in its MC counterpart. On this basis, it was hypothesized that an Fe-Cr alloy with sufficient Cr content will possess much superior oxidation resistance when the structure of the alloy is nanocrystalline (NC) vis-à-vis the same alloy with microcrystalline (MC) structure. On the same basis, it was further hypothesized that with NC structure, considerably lower chromium contents may be sufficient to achieve oxidation resistance similar to common high chromium microcrystalline alloys. These hypotheses were nicely validated by Singh Raman et al. [19,32]. For example, as seen in Figure 5, a NC Fe-10Cr alloy (that was synthesized by mechanical alloying, as described in the preceding section) showed nearly an order of magnitude superior resistance to oxidation than the MC alloy of same composition, during air-oxidation at 300 °C for 3120 min [19,32]. The secondary ion mass spectroscopy (SIMS) depth profiles of Cr through the oxide scale thickness (Figure 6) demonstrate the remarkably higher Cr contents of the protective oxide layer of the NC alloy than that in the MC alloy, thus accounting for the remarkably superior oxidation resistance of the NC alloy [19]. In fact, it took some time (120 min) for sufficient Cr to enrich in the oxide layer to develop a full-fledged protective layer of Cr_2O_3. At this stage, the oxidation rate of NC alloy became insignificant and similar to that for a stainless steel (that possess ~20 wt% Cr), whereas the MC alloy continued to oxidize at a high rate because of its inability to form a full-fledged protective layer of Cr_2O_3. It was possible to explain the stark difference in colour of the NC and MC alloy oxidized for 3120 min (interested readers may refer to [32] for the explanation).

Table 1. Diffusion coefficients of Cr in NC and MC iron [31].

Temperature	Lattice (D_b) m^2/s	GB (D_b) m^2/s	NC Iron (D_n) m^2/s	MC Iron (D_m) m^2/s
300 °C	1.2×10^{-26}	8.6×10^{-22}	1.7×10^{-17}	–
340 °C	3.7×10^{-25}	1.7×10^{-20}	1.6×10^{-16}	–
380 °C	7.6×10^{-24}	2.3×10^{-19}	2.8×10^{-15}	3.6×10^{-19}
840 °C	1.5×10^{-15}	3.7×10^{-12}	–	2.2×10^{-15}

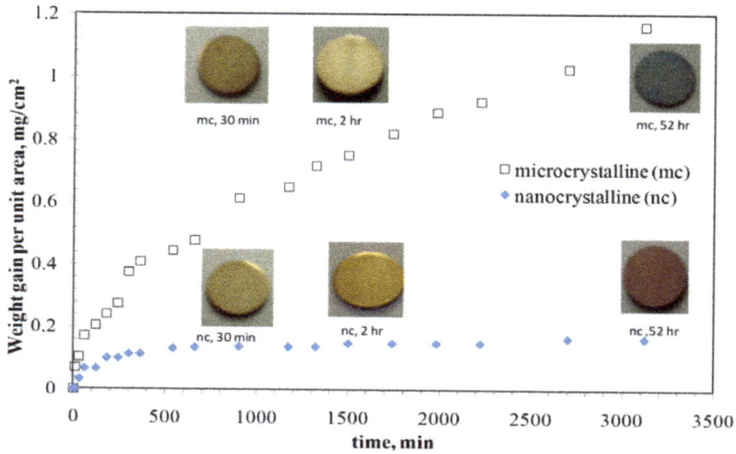

Figure 5. Oxidation kinetics (weight gain versus time plots) of Fe–10Cr alloy with nanocrystalline (nc) and microcrystalline (mc) structures oxidized at 300 °C for 3120 min (insets, show colour of the oxidized samples at different stages of oxidation) ([32], reproduced with copyright clearance).

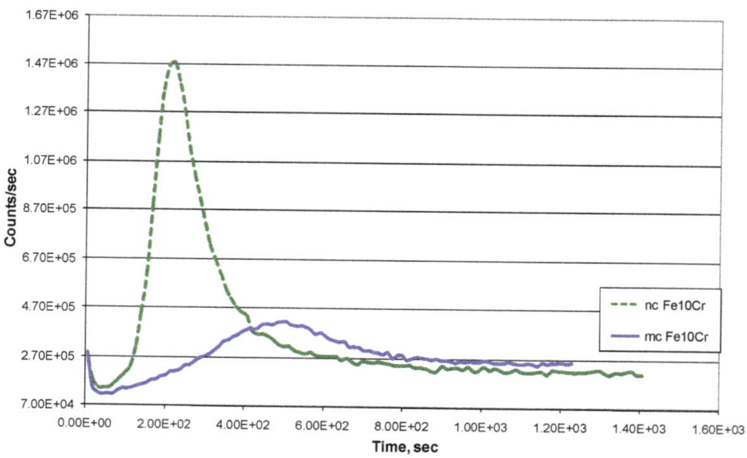

Figure 6. SIMS depth profiles for Cr in the oxide scale of Fe-10%Cr alloy with nanocrystalline (nc or NC) and microcrystalline (mc or MC) structures, oxidized for 120 h in air at 300 °C alloys. ([19], reproduced with copyright clearance).

5. Mechanical Alloying of Fe-Cr-Ni-Zr Alloy Powders and Their Consolidation

As shown in Figure 3, mechanically alloyed Fe-Cr nanocrystalline (NC) powders undergo rapid grain growth when annealed at temperatures above 600–700 °C. Therefore, such powders can lose nanocrystallinity upon exposures to higher temperatures, which means temperatures must be <700 °C for: (a) further thermal processing of such NC powders (such as consolidation/sintering, as described in the preceding section) and (b) applications where the NC structure of alloy is required (such as high temperature oxidation resistant NC alloys). Therefore, it is highly attractive to explore means for retaining nanocrystallinity of mechanically alloyed Fe-Cr alloyed powders at temperatures >700 °C. As described in the preceding section, impurities such as oxide particles pin the grain boundaries and impede grain growth, and as a result, some NC powders of Fe and Fe-Cr alloy resisted grain growth until higher temperature, as shown in Figure 4. Zr readily forms oxides that are thermodynamically stable up to very high temperatures. Following the ball-milling procedure described earlier, Fe, Cr, and Ni powders with and without 2 wt% Zr were mechanically alloyed to synthesize NC Fe-10Cr-5Ni and Fe-10Cr-5Ni-2Zr alloy powders, with an objective of increasing the high temperature processibility of the powders, such as compaction at higher temperature, due to Zr addition [33]. The alloy was added with 5 wt% Ni, for improving ductility of the compacted alloy, since NC alloys are known to suffer from poor ductility [1,2,33]. It should be noted that in the relatively recent times, researchers have successfully employed spark plasma sintering for in-situ consolidation [34].

Grain size of the Fe-10Cr-5Ni alloy powders with and without Zr addition was measured by XRD technique (as described earlier) after annealing for 1 h at various temperatures. As shown in Figure 7, NC powder of Zr-containing alloy resists grain growth during annealing at much higher temperatures. In fact, grain size of Zr-containing NC alloy powder remains well within NC regime upon annealing even at 1000 °C, whereas the Zr-free alloy undergoes rapid grain growth during 1 h annealing at 700 °C and the grains grow into the regime of >100 nm. Mechanically alloyed NC powder of Fe-10Cr-5Ni-2Zr was hot-compacted at 400 °C into discs that could be sintered at a very high temperature (1000 °C), on the basis of the grain growth data shown in Figure 7 [33]. For improving ductility, the NC Fe-10Cr-5Ni-2Zr alloy powder was mixed with an equal amount of microcrystalline (MC) Fe-10Cr-5Ni-2Zr alloy powder and compacted/sintered into discs of Fe-10Cr-5Ni-2Zr alloy with bimodal grain size distribution (50% NC + 50% MC). The alloy with bimodal grain size distribution showed improved ductility [33]. In simple terms, the bimodal alloy has considerable (50%) MC component where dislocations can travel greater distance before being pinned, and hence better ductility.

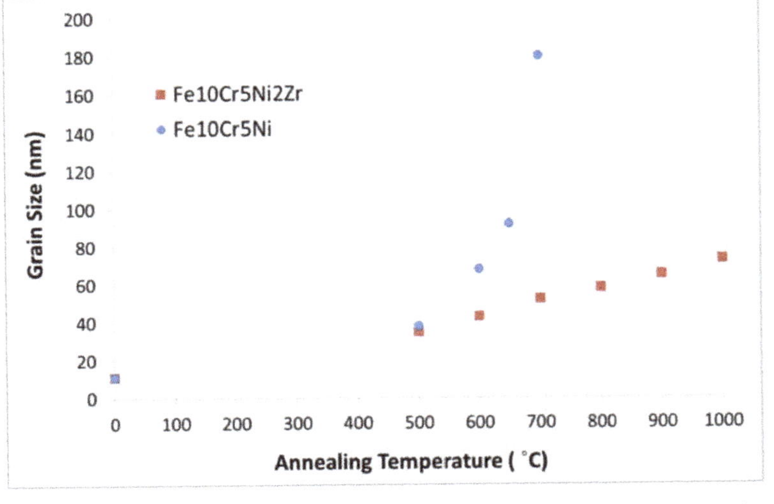

Figure 7. Grain growth of Fe-10Cr-5Ni alloys with and without Zr, after annealing for 1 h at different temperatures [33], reproduced with copyright clearance.

6. Oxidation Resistance of Mechanically Alloyed NC Fe-Cr-Ni-Zr Alloys

Oxidation resistance of NC Fe-Cr alloys has been shown to be considerably superior to that of their MC counterparts [3], as also seen in Figure 5, and the reasons for the superior oxidation resistance of NC alloy is discussed in the literature [32] and summarized in the preceding section. Consistent with this trend, nanocrystalline Fe-10Cr-5Ni-2Zr alloy oxidized at much slower rate than its microcrystalline counterpart, as shown in Figure 8. The role of nanocrystalline structure in improving oxidation resistance is further supported from the observation that the Fe-10Cr-5Ni-2Zr alloy with bimodal grain size distribution (i.e., 50% nanocrystalline and 50% microcrystalline content) oxidized at somewhat higher rate (Figure 8), because of their smaller nanocrystalline content. The most striking observation was that the oxidation resistance of nanocrystalline Fe-10Cr-5Ni-2Zr alloy Cr) was in the same regime as that of a microcrystalline Fe-20Cr-5Ni alloy (whose 20% Cr content is similar to a common stainless steel.) The SIMS depth profiles of Cr through the oxide scale thickness (Figure 9) show considerably lower Cr contents of the protective oxide layer of the microcrystalline Fe-10Cr-5Ni-2Zr alloy, whereas the Cr content was considerably higher in protective oxide of all the alloys with high oxidation resistance [33]. In fact, the relative Cr contents of the protective oxide scales of various alloys commensurate well with their corresponding oxidation resistance.

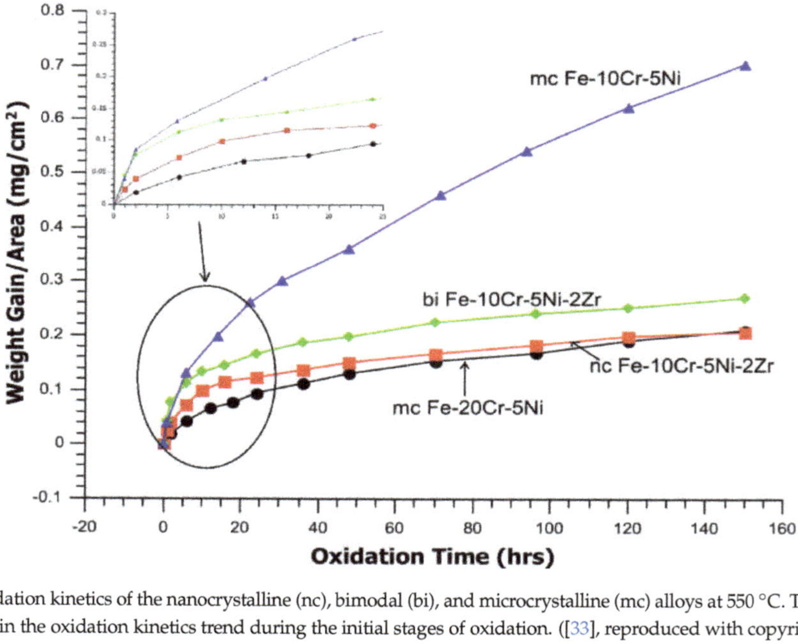

Figure 8. Oxidation kinetics of the nanocrystalline (nc), bimodal (bi), and microcrystalline (mc) alloys at 550 °C. The inset shows the difference in the oxidation kinetics trend during the initial stages of oxidation. ([33], reproduced with copyright clearance).

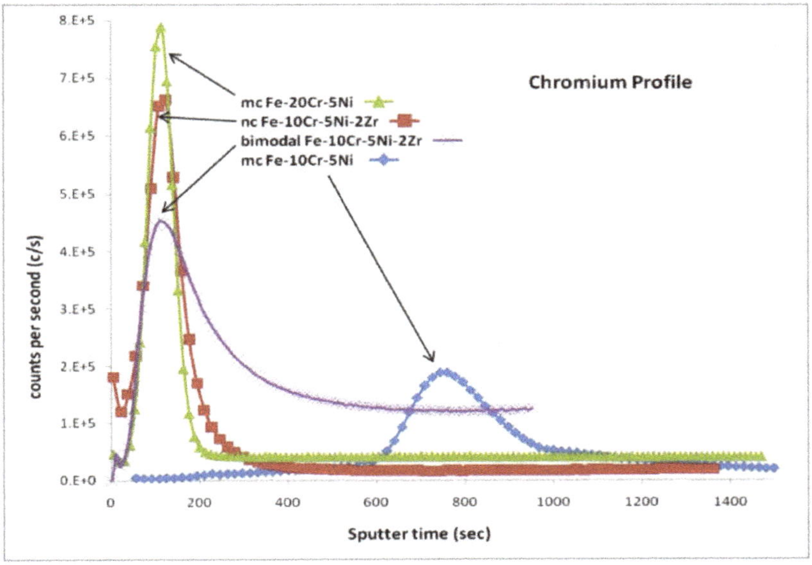

Figure 9. Cr depth profile for oxidised alloys described in Figure 8 (above) measured using secondary ion mass spectroscopy (SIMS). ([33], reproduced with copyright clearance).

7. Opportunities

It is well known [29] that the defect content of Al_2O_3 is considerably less than that of Cr_2O_3, and hence a surface layer of Al_2O_3 is more protective. However, at the Al contents required for developing a continuous protective layer of Al_2O_3 on a steel, the alloy will suffer from unacceptable loss in mechanical property (such as loss of toughness). However, it has been possible to develop Al_2O_3 layer on Fe-Cr-Al and Ni-Cr-Al alloys (that have acceptably low Al contents), with the help of the phenomenon called, "third element effect" [29]. In this phenomenon, the alloy with sufficient Cr first forms a continuous layer of Cr_2O_3, which stops outward diffusion of Al, thereby facilitating Al accumulation (just underneath the Cr_2O_3 layer) to the level that is sufficient for the development of a full-fledged layer of Al_2O_3. Thus, with the "third element effect" of Cr, the alloy develops a protective layer of Al_2O_3 in spite of the low Al content of the alloy. As discussed in the preceding sections, NC structure profoundly accelerates diffusion; hence, it is expected to facilitate the "third element effect". Therefore, it will be interesting to investigate the role of NC structure of mechanically alloyed systems in promoting the "third element effect".

8. Conclusions

Nanocrystalline (NC) powders of Fe-Cr alloys can be synthesized by mechanical alloying. However, consolidation of these powders into monolithic mass is a challenge because of their high hardness due to body-centered cubic crystalline structure of Fe-Cr systems. In order to promote plastic flow, compaction is required to be carried out at high temperatures; however, too high a temperature can cause grain growth and loss of NC structure. A temperature-time window is optimized for hot-compaction of NC powders of Fe-Cr alloys without losing NC structure. In order to enable hot processing of the powders at much higher temperatures without losing NC structure, Zr was added to the alloy during mechanical alloying that pins grain boundaries and suppresses grain growth. As a result, compaction temperature of powders of Fe-10Cr-5Ni-2Zr alloy can be as high as 1000 °C.

Oxidation resistance of compacted nanocrystalline (NC) Fe-10Cr alloy is found to be remarkably superior to its microcrystalline (MC) counterpart. The superior oxidation resistance of NC alloy is attributed to the much greater grain boundary volume of NC structure that causes sufficient diffusion of Cr in oxide layer and development of a protective layer of Cr-oxide, whereas the MC alloy fails to develop this protective layer because of insufficient Cr diffusion. In fact, Cr content of the protective oxide layer developed during oxidation of NC Fe-10Cr-5Ni-2Zr alloy at 550 °C is similar to that of a MC Fe-20Cr alloy (i.e., common stainless steel); hence, the two alloys show similar oxidation resistance. Regarding further opportunities, it will be interesting to investigate the role of NC structure of mechanically alloyed systems in promoting the "third element effect".

Funding: This research received no external funding.

Institutional Review Board Statement: Not applicable.

Informed Consent Statement: Not applicable.

Data Availability Statement: Data is contained within the article.

Conflicts of Interest: The author declares no conflict of interest.

References

1. Kumar, K.; Van Swygenhoven, H.; Suresh, S. Mechanical behavior of nanocrystalline metals and alloys. *Acta Mater.* **2003**, *51*, 5743–5774. [CrossRef]
2. Suryanarayana, C.; Froes, F. The structure and mechanical properties of metallic nanocrystals. *Metall. Trans. A* **1992**, *23*, 1071–1081. [CrossRef]
3. Mahesh, B.V.; Raman, R.K.S. Role of Nanostructure in Electrochemical Corrosion and High Temperature Oxidation: A Review. *Metall. Mater. Trans. A* **2014**, *45*, 5799–5822. [CrossRef]
4. Koch, C.C.; Youssef, K.M.; Scattergood, R.O.; Murty, K.L. Breakthroughs in Optimization of Mechanical Properties of Nanostructured Metals and Alloys. *Adv. Eng. Mater.* **2005**, *7*, 787–794. [CrossRef]
5. Rofagha, R.; Langer, R.; El-Sherik, A.M.; Erb, U.; Palumbo, G.; Aust, K.T. The corrosion behaviour of nanocrystalline nickel. *Scripta Metall. Mater.* **1991**, *25*, 2867. [CrossRef]
6. Rofagha, R.; Erb, U.; Ostrander, D.; Palumbo, G.; Aust, K.T. The effects of grain size and phosphorus on the corrosion of nanocrystalline Ni-P alloys. *Nanostruct. Mater.* **1993**, *2*, 1–10.
7. Karimpoor, A.A.; Erb, U.; Aust, K.T.; Palumbo, G. High strength nanocrystalline cobalt with high tensile ductility. *Scripta Mater.* **2003**, *49*, 651–656. [CrossRef]
8. Cheng, S.; Ma, E.; Wang, Y.M.; Kecskes, L.J.; Youssef, K.M.; Koch, C.C.; Trociewitz, U.P.; Han, K. Tensile properties of in situ consolidated nanocrystalline Cu. *Acta Mater.* **2005**, *53*, 1521. [CrossRef]
9. Gleiter, H. Nanocrystalline materials. *Prog. Mater. Sci.* **1989**, *33*, 223.
10. Trapp, S.; Limbach, C.T.; Gonser, U.; Campbell, S.J.; Gleiter, H. Enhanced Compressibility and Pressure-Induced Structural Changes of Nanocrystalline Iron: In Situ Mössbauer Spectroscopy. *Phys. Rev. Lett.* **1995**, *75*, 3760. [CrossRef] [PubMed]
11. Roy, R.A.; Roy, R. Diphasic xerogels: I. Ceramic-metal composites. *Mater. Res. Bull.* **1984**, *19*, 169–177. [CrossRef]
12. Seo, J.H.; Kim, J.K.; Yim, T.H.; Park, Y.B. Textures and Grain Growth in Nanocrystalline Fe-Ni Alloys. *Mater. Sci. Forum* **2005**, *475–479*, 3483–3488. [CrossRef]
13. Hibbard, G.D.; Aust, K.T.; Erb, U. Thermal stability of electrodeposited nanocrystalline Ni–Co alloys. *Mater. Sci. Eng. A* **2006**, *433*, 195–202. [CrossRef]
14. Groza, J.R. Nanocrystalline Powder Consolidation Methods. In *Nanostructured Materials: Processing, Properties, and Applications*; Koch, C.C., Ed.; William Andrew Pub.: Norwich, NY, USA, 2007.
15. Rawers, F.; Biancaniello, R.; Jiggets, R.; Fields, R.; Williams, M. Warm-hip compaction of attrition-milled iron alloy powders. *Scr. Mater.* **1999**, *40*, 277. [CrossRef]
16. Youssef, K.M.; Scattergood, R.O.; Murty, K.L.; Koch, C.C. Ultratough nanocrystalline copper with a narrow grain size distribution. *Appl. Phys. Lett.* **2004**, *85*, 929. [CrossRef]
17. Elkedim, O.; Cao, H.S.; Guay, D. Preparation and corrosion behavior of nanocrystalline iron gradient materials produced by powder processing. *J. Mater. Process. Technol.* **2002**, *121*, 383. [CrossRef]
18. Guruswamy, S.; Loveless, M.R.; Srisukhumbowornchai, N.; McCarter, M.K.; Teter, J.P. Processing of Terfenol-D alloy based magnetostrictive composites by dynamic compaction. *IEEE Trans.* **2000**, *36*, 3219–3222. [CrossRef]
19. Raman, R.K.S.; Gupta, R.K. Oxidation Resistance of Nanocrystalline vis-à-vis Microcrystalline Fe-Cr Alloys. *Corros. Sci.* **2009**, *51*, 316–321. [CrossRef]
20. Klug, H.P.; Alexander, L. *X-ray Diffraction Procedures for Polycrystalline and Amorphous Materials*, 2nd ed.; Wiley-VCH: Weinheim, Germany, 1974.

21. Gupta, R.K. Synthesis and Corrosion Behavior of Nanocrystalline Fe-Cr Alloys. Ph.D. Thesis, Monash University, Melbourne, Australia, 6 May 2010.
22. Siegel, R.W. Mechanical properties of nanophase materials. *Mater. Sci. Forum* **1997**, *235–238*, 851.
23. Malow, T.A.; Koch, C.C. Grain growth in nanocrystalline iron prepared by mechanical attrition. *Acta Mater.* **1997**, *45*, 2177. [CrossRef]
24. Gupta, R.K.; Raman, R.K.S.; Koch, C.C. Grain Growth Behaviour and Consolidation of Ball Milled Nanocrystalline Fe-10Cr Alloy. *Mater. Sci. Eng. A* **2008**, *494*, 253–256. [CrossRef]
25. Moelle, C.H.; Fecht, H.J. Thermal stability of nanocrystalline iron prepared by mechanical attrition. *NanoStructured Mater.* **1995**, *6*, 421. [CrossRef]
26. Bonetti, E.; del Bianco, L.; Pasquini, L.; Sampaolesi, E. Thermal evolution of ball milled nanocrystalline iron. *NanoStructured Mater.* **1999**, *12*, 685. [CrossRef]
27. Natter, H.; Schmelzer, M.; Loeffler, M.S.; Hempelmann, R. In-situ X-ray crystallite growth study on nanocrystalline Fe. *J. Metastable Nanocryst. Mater.* **2000**, *8*, 683–688. [CrossRef]
28. Perez, R.J.; Jiang, H.G.; Lavernia, E.J. Grain size stability of nanocrystalline cryomilled Fe-3wt.% Al alloy. *NanoStructured Mater.* **1997**, *9*, 71. [CrossRef]
29. Kofstad, P. *High Temperature Corrosion*; Elsevier Applied Science and Publishers Ltd.: New York, NY, USA, 1988.
30. Wagner, C. Oxidation of alloys involving noble metals. *J. Electrochem. Soc.* **1952**, *99*, 103. [CrossRef]
31. Wang, Z.B.; Tao, N.R.; Tong, W.P.; Lu, J.; Lu, K. Diffusion of chromium in nanocrystalline iron produced by means of surface mechanical attrition treatment. *Acta Mater.* **2003**, *51*, 4319. [CrossRef]
32. Raman, R.K.S.; Gupta, R.K.; Koch, C.C. Synthesis Challenges and Extraordinary Resistance to Environmental Degradation of Nanocrystalline vis-à-vis Microcrystalline Fe-Cr Alloys. *Philos. Mag.* **2010**, *90*, 3233–3260. [CrossRef]
33. Mahesh, B.V.; Raman, R.K.S.; Koch, C.C. Bimodal grain size distribution: An effective approach for improving the mechanical and corrosion properties of Fe-Cr-Ni alloys. *J. Mater. Sci.* **2012**, *47*, 7735–7743. [CrossRef]
34. Kumar, R.; Bakshi, S.R.; Joardar, J.; Parida, S.; Raja, V.S.; Raman, R.K.S. Structural Evolution During Milling, Annealing and Rapid Consolidation of Nanocrystalline Fe-10Cr-3Al Powder. *Materials* **2017**, *10*, 272. [CrossRef] [PubMed]

MDPI
St. Alban-Anlage 66
4052 Basel
Switzerland
Tel. +41 61 683 77 34
Fax +41 61 302 89 18
www.mdpi.com

Metals Editorial Office
E-mail: metals@mdpi.com
www.mdpi.com/journal/metals

www.ingramcontent.com/pod-product-compliance
Lightning Source LLC
Chambersburg PA
CBHW062059220926
51850CB00010B/2819
9783036521176